SPACE INVADERS

SPACE INVADERS

HOW ROBOTIC SPACECRAFT EXPLORE THE SOLAR SYSTEM

Michel van Pelt

Copernicus Books
An Imprint of Springer Science+Business Media

In Association with
Praxis Publishing Ltd

Softcover re-print of the Hardcover 1st edition 2007

Published in the United States by Copernicus Books, an imprint of Springer Science+Business Media.

Copernicus Books
Springer Science+Business Media
233 Spring Street
New York, NY 10013
www.springer.com

9 8 7 6 5 4 3 2 1

ISBN-10: 1-4899-9012-7
ISBN-13: 978-1-4899-9012-9

To Stefania,
who flies me to the Moon and
lets me play among the stars

CONTENTS

PREFACE

Contrary to popular belief, Neil Armstrong and Buzz Aldrin were not the first to land on the Moon. It was Luna 2, a robotic space probe the size of a large beach ball, which arrived there 10 years earlier. First to reach another planet, first to land on Mars and Venus, first to touch an asteroid, first to see the outer planets close up, first to land on the moon of another planet; the list of robotic space achievements is long and impressive. In fact, robots ventured into space even before robot assemblers took over the car factories and robot vacuum cleaners and robodogs were let loose in our houses.

Until the 1950s, before the start of the Space Age, space exploration was envisioned to become the domain of bold heroes in spacesuits, shooting through the void in elegant rocketships. Buck Rogers and his colleagues were the public's stereotypes of space explorers, and even more scientific books and artworks focused on manned capsules, spaceplanes, space stations, and planetary bases. No one seemed to foresee that the exploration of space would actually become the domain of unmanned machines. At a fraction of the costs of manned spacecraft, automatic probes became the true explorers of the Solar System.

Although the manned space programs get most of the media attention, robotic space probes are still going further and faster than any astronaut ever has. These machines do not need oxygen, food or sleep, can spend decades in high vacuum, and can endure extreme temperatures without complaining. Unlike people, they can be made small and are adaptable to nearly any hostile environment.

When uncrewed probes are lost, the tragedy is much less grave than it would have been had astronauts been on board. Space robots can be replaced, but people cannot. Moreover, space probes do not have to come back after completion of their mission.

Our hardy robots have gone inside the turbulent atmosphere of the giant gas planet Jupiter, moved out beyond the orbit of Pluto and landed on asteroids while we humans are still fiddling around in low Earth orbit. True, we have set foot on the Moon, but the last time we did that was well over 30 years ago. However, even while you are reading this, brave machines are trailblazing the Solar System in preparation for future human exploration. This book tries to give these courageous satellites and probes the glory they deserve.

A way to understand unmanned spacecraft is to compare them with living beings. Like animals and people, spacecraft need energy, interact with their environment, and have different body parts (called "sub-systems") that perform necessary functions. Similar to many organisms, space probes need to keep themselves within a certain temperature range, have a skeletal structure, a brain to process sensory inputs and regulate internal functions, and equipment to see, smell, hear, feel, and communicate. In fact, we regularly react to less complicated machines such as cars and computers as if they were alive, especially when they do not work properly (the number of people who physically assault their computer monitor is quite high . . .).

In this book, comparisons between spacecraft, everyday machines, and animals are used to explain complicated technical issues. You will see that you do not need to be a rocket scientist to understand the basic workings of a spacecraft.

In the Introduction we look at the history of robotic space exploration, and how astronauts and robots are complementary in pushing the frontiers of human knowledge. We will see how a robotic space mission is started with new ideas, scientific questions and a lot of engineering, and are then going to dissect a space probe to see how it works. We will move on to find out how spacecraft are tested and qualified for a trip into interplanetary space. The topic of the following chapter is scientific instruments – the many eyes through which a robotic space explorer investigates new worlds.

As you read on, you will discover how these ingenious automatic probes are launched, and then how they have pushed the boundaries of our imagination and scientific understanding. The knowledge they provide lives on, even after the last spark of energy has been squeezed out of their batteries. They are truly paving the way for us.

I want to thank my girlfriend Stefania for all her support and suggestions. Alessandro Atzei, Peter Buist, Dennis Gerrits, Zeina Mounzer and Ron Noteborn, once again loads of gratitude for checking the engineering details and spelling, as well as offering important facts and fictions that certainly enriched this book. They made sure my descriptions remained precise and kept me from becoming too much of a fuzzy dreamer. I also thank Arno Wielders for helping to make the chapter on payload instruments quite a bit more understandable and complete, and Stella Tkatchova has my gratitude for reading the text from a non-engineering perspective – something nearly impossible to do for the techno-biased rest of us. As before, I owe a lot to Clive Horwood of Praxis and to Paul Farrell and my good friend Harry Blom of Copernicus for their confidence and support. John Mason of Praxis thoroughly scrutinized the first draft of this book and significantly improved both the style and the contents; any mistakes that remain are solely my responsibility. Finally, I wish to thank Praxis copy editor Alex Whyte for converting my "Eurenglish" into American.

It's human nature to stretch, to go,
to see, to understand. Exploration is not a choice, really;
it's an imperative.

Michael Collins, Apollo 11 astronaut

Beep, beep, beep.

Sputnik 1's famous first message from Earth orbit

1

INTRODUCTION

IN our modern high-tech world we very much depend on the unique benefits of space, although we often do not realize it. Satellites broadcast a television news program to us, including a report from Moscow sent to the news studio by a journalist via a satellite link. The report concerns the latest international agreements to limit the production of CFC-chemicals, which, when they escape into the atmosphere, can destroy the ozone layer. Earth-observing satellites helped to reveal the extent of the problem over Antarctica, and detected the smaller ozone holes over the Arctic. This new knowledge convinced us that we must prevent further damage to this precious atmospheric shield, which protects us against the harmful ultraviolet radiation of the Sun. Without continuous monitoring from space, it would have been impossible for scientists and politicians to appreciate the scale of the ozone problem. Without urgent action, the ozone holes could have grown much larger, considerably increasing our chances of developing skin cancer or cataracts.

Another news item is about the International Space Station, where astronauts are investigating the effects on the human body in microgravity conditions. This will help us to plan and design future crewed expeditions to the Moon and Mars. The astronauts were recently joined by a so-called Flight Participant – a rich person who bought a ticket on board a Russian Soyuz spacecraft to visit the space station. Thanks to space technology, it is now possible for non-professional astronauts to experience the thrill of a launch, weightlessness, and the magnificent view of the Earth from orbit. It is still very expensive, but ticket prices may come down in the future and allow more and more people to take their holiday of a lifetime.

The weather forecast following the news is based on pictures made from orbit by meteorological satellites. We pick up the phone and within seconds, using satellite telephone links, we talk to friends and family on the other side of the planet. We learn that an adventurous relative narrowly escaped drowning when his yacht sank in a storm in the Pacific Ocean. He was rescued by a Navy ship that received a distress and location signal from a transmitter on his life raft. The signal had been forwarded to them by a Search and Rescue radio system on board a satellite. We browse the Internet and, via an orbital connection, check how the stock exchange is doing in Tokyo. Next, we go to visit a colleague who wants to show his new house, which has advanced solar arrays on the roof that are based on solar cells developed for spacecraft. The navigation system in our car accurately guides us to his place using GPS signals from positioning system satellites.

Meanwhile, we are curious about the endless Universe around us and send space probes to the planets and beyond. We are discovering the magnificence and the unbelievable diversities of the new worlds we explore, and at the same time start to understand how unique our own Earth really is. The exploration of the Solar System is a marvelous, inspiring adventure, and the science, the technology and the challenges involved make it an endeavor like no other.

The quest for scientific knowledge is a worthy goal in its own right, but the exploration of the Solar System has direct uses as well. For instance, what the investigation of the planets tell us about the runaway greenhouse effect on Venus and the thin, ozone-less atmosphere of Mars helps us to make better predictions about what is happening to our own atmosphere here on Earth.

We are used to all these space applications, which have been seamlessly integrated into our daily lives to such an extent that we hardly think about them any more. However, until 50 years ago this was all pure science fiction; there were simply no satellites and no interplanetary probes flying through the Solar System.

FROM THE GROUND UP

The sky at night has fascinated us for as long as we can look back. Early man thought that the Earth was the center of everything, and that the stars were mere points of light hanging overhead. Most stars appeared to be fixed forever in the same patterns or constellations, and by drawing lines between stars appearing close together we could picture animals and gods in them.

A few stars, however, moved differently around the sky, and the ancient Greek called these objects "planets," meaning "wandering stars." At first we thought the Sun, Moon and planets all revolved around us, until Copernicus revived a long forgotten idea of the Greek philosopher Aristarchus of Samos that the Sun was actually the queen of the Solar System and that the planets, including Earth, were orbiting her. The movements in the sky made much more sense this way.

Planets nevertheless remained mysterious specs of light until the telescope was invented. Galileo Galilei used the new instrument to discover that the Moon was scarred by thousands of large craters, and that the planet Jupiter had four large moons of its own. As telescope technology advanced, the amazing rings of Saturn were observed, and the outer planets Uranus, Neptune and Trans-Neptunian objects such as Pluto were discovered. (It now appears that there are many more small Pluto-type planets beyond Neptune.)

We discovered that the Sun is only one of 200 billion stars in a disk-shaped system we see edge-on as the Milky Way, and that there are an uncountable number of other galaxies. The planets proved to be more than points of light; Venus was found to have a thick atmosphere, Mars had polar caps, and Jupiter was surrounded by ever-changing bands of clouds.

Telescopes grew larger and larger, but we always had to remain on Earth and could never actually visit the planets to inspect them close-up. Because of this limitation, misconceptions arose that could not be proven false, such as the idea that the craters on the Moon had a volcanic origin (while they are in fact created by the impacts of comets, asteroids and meteorites), or that Mars was covered with long, possibly artificial, channels or canals and intriguing changing patterns of vegetation (the canals later proved to be an optical illusion of vague features together appearing as lines, and the changing spots of vegetation were shown to be dark rock plateaus that are sometimes partly covered by lighter colored dust). However, in the mid-twentieth century the state of technology offered us a way to visit these mysterious places – in person or using remotely operated automatic probes.

Chapter 1

Rockets and Satellites

The history of satellites and space probes is directly linked to the history of the rocket, as this was the only machine that could achieve the extreme velocities and altitudes required to put objects into space.

The first to really understand that we would need to develop large rockets before we could even think of developing artificial satellites was a Russian schoolteacher by the name of Konstantin Tsiolkovsky (1857–1935). Born deaf, he was unable to attend normal elementary schools, but he taught himself physics and mathematics from books. His father sent him to Moscow to study further, but limited by his deafness he spent most of his time in the great Moscow libraries. Nevertheless, just as he had done before, he educated himself using the library books he found, and in 1882 he managed to get a job as a mathematics teacher in the town of Kaloega, south of Moscow.

Ever since Tsiolkovsky was 12 years old, he had been fascinated by rockets and their possibilities for the exploration of space. He understood that only rockets can still work in the vacuum that exists above the Earth's atmosphere; unlike, for instance, propeller engines, they do not need air because rockets produce their own gasses to propel themselves. Furthermore, unlike car engines and jet engines, rockets do not depend on atmospheric oxygen to sustain combustion since they carry their own oxidizers.

As the fuel and oxidizer (oxygen) mix in the combustion chamber and ignite, they form a fast expanding gas that is expelled at great velocity through the rocket nozzle. Following the "Action = Reaction" principle of Newton, the rocket itself will be thrust forward (the reaction) in the opposite direction to that in which the exhaust gases are going (the action). It is like standing on a skateboard and throwing rocks in the backward direction, which makes you move forward.

In his free time, Tsiolkovsky increasingly busied himself with the theories of spaceflight. He wrote about steerable airships, manned rockets, self-sustaining space stations and spacesuits. Since even motorized airplanes were still science fiction at the time, his fellow citizens considered him to be rather eccentric. However, Tsiolkovsky was far from mad. He understood the physics of rocketry and formulated the basic mathematical equations describing the workings of a rocket and the dynamics of its flight.

In 1903, the same year that the Wright brothers' first motorized plane took off, he invented a method of "staging": stacking series of rockets on top of each other. This method increases the efficiency of launchers, as

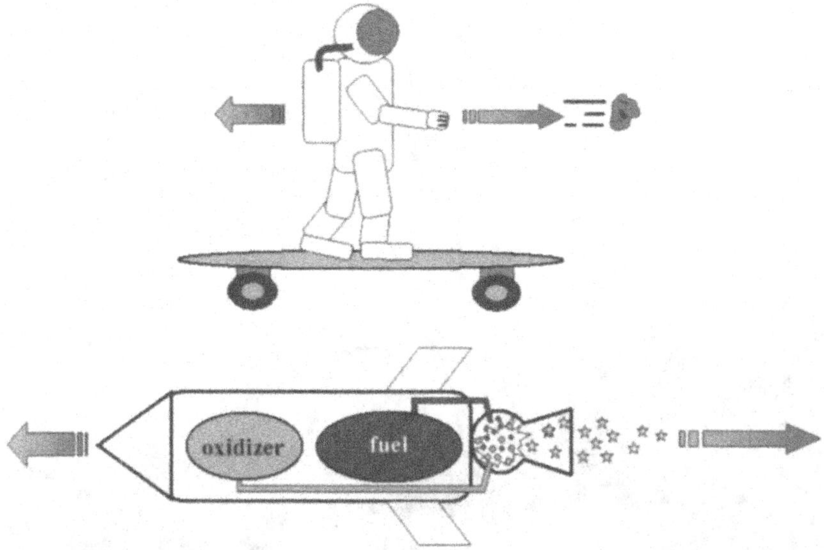

FIGURE 1.1 Both someone on a skateboard throwing rocks and a rocket creating thrust by burning oxidizer and fuel move forward because of Newton's Action = Reaction principle.

stages can be dropped off when their tanks run dry, ensuring that their useless empty mass does not have to be dragged all the way into orbit.

Under the Russian Tsar's regime Tsiolkovsky never enjoyed any government support, but this changed dramatically after the 1917 Revolution and the Soviets came to power. In 1919 he was even made a Member of the prestigious Academy of Sciences and between 1925 and 1932 no less than 60 of his papers were published.

Tsiolkovsky died in 1935, and although he never built any rockets himself, he inspired a number of young Russian rocket experimenters who, in the 1930s, began to build small experimental rockets based on liquid propellants. As a testimony to his vision, Tsiolkovsky's most famous words can be found on his gravestone: "The Earth is the cradle of mankind, but one cannot stay in the cradle forever."

Also in other countries scientists and engineers started to think about the possibilities for rockets to reach outer space. In 1918 American Robert Goddard had already built and launched an extensive series of small rockets, of which the largest measured 1.70 meters (5.1 feet) in length and weighed 20 kilograms (44 pounds). They were supposed to lead to rocket weapons that could be used in the First World War, but the fighting was over before Goddard's developments were finished and the military lost interest in his work.

Nevertheless, Goddard continued his research and experimentation,

FIGURE 1.2 Dr Robert H. Goddard with the world's first liquid propellant rocket, which he launched in 1926. Unlike in modern rockets, the nozzle is on top. [NASA]

and the following year he published a report called "A Method of Reaching Extreme Altitudes" in which he described how a rocket could reach the Moon and signal its arrival by use of flash powder igniting at impact. However, in spite of the scientifically sound basis of his proposals, Goddard's ideas were met with disbelief and ridicule.

Persevering, in 1926 he flew the first ever liquid-fueled rocket. Like Tsiolkovsky, he realized that rocket systems based on liquid propellant have a number of advantages over the simpler solid propellant (gunpowder) rockets used then for artillery and fireworks. Liquid propellant rockets have a lower mass relative to their thrust and, unlike solid propellant rockets, can be throttled up or down, or stopped entirely.

In 1935 Goddard's rockets had broken the sound barrier and reached a height of 1.5 kilometers (0.9 mile), but the US government did not seem to appreciate the possibilities of this technology. During the Second World War all they did was assign Goddard a contract for the development of a starting rocket to help carrier-based aircraft to take off from their ships.

In the meantime Germany had seriously started to develop rockets. In 1923 Hermann Oberth published his classic book *Die Rakete zu den Planetenräumen* ("The Rocket into Planetary Space"), stimulating young German engineers to experiment with rockets themselves. Oberth's own interest in rockets had been stimulated when, at the age of 12, his mother had given him a copy of Jules Verne's *From the Earth to the Moon.*

Oberth's inspiring book resulted in the launch of the first ever rocket-powered aircraft in 1928, while in the same year car manufacturer Fritz von Opel achieved 200 kilometers (124 miles) per hour in a rocket car and reached 153 kilometers (95 miles) per hour flying a glider fitted with solid propellant rockets. In 1931, Germany's first liquid propellant rocket left a proving ground near Berlin, while two years later Viennese Professor Eugene Sänger published a book called *Raketenflugtechnik* ("Rocketflight technology") that contained a concept for a high-speed rocket-propelled research aircraft.

Unlike in America, the developments in rocketry caught the serious attention of the military. German Army leaders saw the development of large, powerful rocket bombs as a way to circumvent the ban on the use of large cannons, as stipulated in the Treaty of Versailles, which Germany had been forced to sign after its defeat in the First World War.

The futuristic engineers were recruited by the military, and instead of reaching for space they found themselves developing ballistic missiles. However, both parties needed the same rocket technology to achieve their goals, and while the engineers had the brains, the military could provide the much-needed financial and logistical support.

Wernher von Braun, one of the group's younger spaceflight enthusiasts, greatly impressed the military leaders. Although he was only 20 years old, they appointed him to lead the technical part of their extensive rocket development.

Soon the development team's expanding activities required a dedicated development, production and launch center, and it was decided to create such a center in Peenemünde, a remote, sparsely inhabited place at the

Baltic Sea in the north of Germany, far from spying Allied eyes. Laboratories, wind tunnels, test stands, launch platforms and housing facilities for the 2,000 rocket scientists and 4,000 supporting workers, plus their families, were built there.

Until the Second World War rockets had been small and incapable of carrying much or going very high, but with the support of the Nazi war machine von Braun's team developed a 14-meter-high (46-foot-high) monster capable of launching a 738-kilogram (1,630-pound) warhead a distance of 418 kilometers (260 miles). In 1942 they launched their first rocket into space, reaching a height of 80 kilometers (50 miles), a distance of 193 kilometers (120 miles), and attaining a velocity of over 5,300 kilometers (3,300 miles) per hour.

The A4 rocket that the Peenemünde group created was soon renamed "Vengeance 2" by the Nazi government. Before the war was over, thousands of V2 rocket bombs were launched at London and liberated Antwerp, blowing away entire blocks of houses in an instant.

Fortunately, the V2 and the other "Wonder Weapons" of Nazi Germany were unable to turn the tide of the war, and in 1945 Germany

FIGURE 1.3 A Bumper V2, based on the German V2 rocket of the Second World War, leaves Cape Canaveral in 1950. [NASA]

surrendered. Understanding that ballistic missiles were going to play a major role in military power, a race developed among the Soviet, British and American military forces to find as many German rocket engineers and blueprints, and as much hardware, as they could get their hands on.

Located in what was soon to become East Germany, Peenemünde was destined to be under Soviet control. However, von Braun and most of the leading figures of the Peenemünde team had fled the rocket base, carrying as much documentation and hardware with them as they could. Convinced that their best chance for continuing rocket development lay with the United States, they surrendered to American forces.

They also gave them directions to find crucial documentation that they had hidden in an underground mine. Some 14 tons of design blueprints, test reports and other archive materials were transported to the American sector just before the British took control of the region, as agreed. The material would have become British property had it remained in the mine.

In another daring smuggling operation, the American Army managed to transport 341 rail carts of rocket equipment out of an underground factory in the Harz mountains, just before the area had to be turned over to the Soviets.

The Americans now had the best cards for the future: the rockets, the documentation and most of the German experts. Nevertheless, the Russians also managed to retain a few rocket experts and some rocket parts, while the British obtained some hardware. (Some of the V2s they "liberated" can still be seen in the Imperial War Museum and the Science Museum in London.)

America and Russia soon moved the German engineers to their own development centers and put them to work on new ballistic missile projects. With the Cold War just starting, the development of large rockets became a top priority in America and Russia. Once again, von Braun and his consorts found themselves working on military weapons instead of spacecraft.

In Russia, chief designer Sergei Korolev was the driving force behind the Soviet rocket program. Originally a carpenter and slater turned aircraft designer, Korolev decided to pursue spaceflight after having met Tsiolkovsky. He started equipping gliders and other airplanes with rocket engines and even flew some of them himself.

In the 1930s Korolev grouped with a couple of other rocket fanatics and started GIRD, a Russian abbreviation for "Group for Research of Reactive Propulsion." Working without any financial support, they jokingly explained that their club's name meant "Group of Engineers Working without Money."

Nevertheless, in 1933 they managed to launch the first Soviet liquid propellant rocket, designed by rocket pioneer Fridrikh Tsander, to an altitude of 400 meters (1,300 feet). Now the military became interested, offered financial support and even founded a special institute for the development of rocketry.

However, the Soviet Union under Stalin was a dangerous place in which to live and work. Together with almost half the officer corps of the Red Army, the Field Marshal who supported Korolev and his colleagues was accused of conspiring with Nazi Germany and executed on Stalin's orders. Because of his link with the Marshal, even Korolev was arrested. According to the accusations, he was supposed to have tried to sabotage a rocketplane he was working on. Korolev narrowly escaped being shot, but was sentenced to 10 years of forced labor in the infamous Kolyma gold mines of Siberia.

However, when the Second World War broke out Stalin realized that he desperately needed technicians and Korolev was brought back to Moscow. Still under custody, he was put to work on military planes and rockets. After the war, Korolev continued his work on rocketry together with the captured German engineers and scientists of von Braun's Peenemünde team.

Taking advantage of the German experts' knowledge, Korolev and his group were soon building Russian copies of the V2 rocket. Unlike their colleagues in America, the German experts in Russia were dismissed and sent home after their V2 work was completed. Korolev's institute, now fully supported by Stalin, continued its work and started to launch rockets that were increasingly larger and more advanced. Soon they were capable of launching satellites in Earth orbit.

To put a satellite into orbit, you need to give it a very high velocity. Imagine a tower with a cannon on top of it, aimed horizontally to the horizon. The fired shell will be given a velocity in the horizontal direction, but as soon as it leaves the cannon's muzzle, gravity will start pulling it down. As a result the shell's trajectory is curved downward, and it hits the ground a certain distance from the cannon.

If you manage to put more gunpowder into your cannon and fire it again, the second shell will get a higher initial velocity, and therefore fly further before it impacts. Now imagine a truly huge gun, able to give a projectile the enormous velocity of 7.5 kilometers (4.7 miles) per second. Again, as soon as the shell is fired, gravity starts pulling it down into a curved trajectory. However, the shell now manages to fly very, very far in the horizontal direction before it falls to ground level. In fact, at this velocity it flies over the horizon and, as it falls down, the ground level also moves down because of the Earth's curvature.

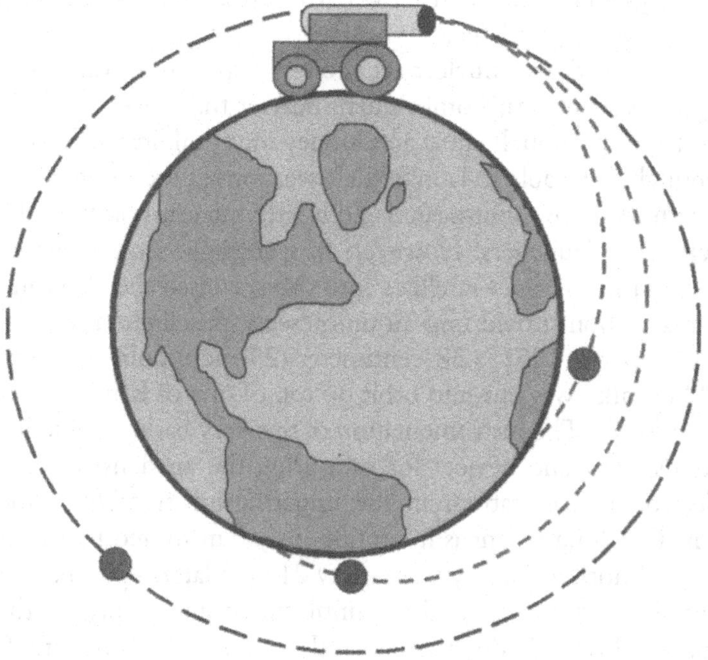

FIGURE 1.4 The faster a cannonball, the further it will fly. If shot at a high enough velocity and at sufficient altitude, the projectile could go into orbit.

At 7.8 kilometers (4.8 miles) per second the curvature of the projectile's trajectory is exactly the same as the curvature of the Earth. The shell thus never reaches the ground, but falls all the way around the Earth to hit the cannon in the back. If you could quickly remove the cannon, the shell would continue to fall around the planet in a circle. It would then be in Earth orbit and have become a satellite.

In reality this does not work, owing to the atmospheric drag that continuously slows down the projectile. To keep the projectile in orbit you would need to put it above the atmosphere, at least to an altitude of about 200 kilometers (130 miles).

Satellites are not launched by cannons, mostly because the enormous acceleration needed to bring them from 0 to 7.8 kilometers (4.8 miles) per second over the length of the cannon's barrel would be far too high for most on board equipment to survive.

Instead, rocket launchers are used to relatively slowly increase the velocity and bring the satellite above the atmosphere. This requires an enormous amount of propellant and supporting equipment. Typically, it takes 100 kilograms (230 pounds) of propellant and launcher hardware to place 1 kilogram of satellite in a low Earth orbit. Hence a satellite launcher needs to be very, very large.

In the 1950s Soviet nuclear technology was behind that of America, making Russia's nuclear bombs much heavier than those of its rival. The Soviets had to put much effort and money into building rockets that were large enough to be able to launch their weapons at the United States. The USA, with its more sophisticated and lighter nuclear bombs, only needed relatively small launchers. However, in spaceflight large launchers mean that you can put heavier satellites into orbit, and so the Russian military disadvantage soon turned into an unforeseen spaceflight advantage.

On October 4, 1957, a 58-centimeter (23-inch) polished metal sphere called "Sputnik" was put into orbit on top of one of Korolev's converted ballistic missiles. The instrumentation of this very basic satellite included a radio transmitter and devices for measuring the air density, temperature and electron concentrations in the uppermost part of the atmosphere. Using its four long antennas it continuously transmitted its coded "beep, beep" signal until its battery was empty 21 days later. Sputnik 1 remained in orbit for a total of 96 days, until the minute drag of the upper atmosphere slowed it down sufficiently to be pulled back to Earth. It burned up in the atmosphere after making 1,400 orbits around our planet.

The attention of the world for the Soviet success was overwhelming, enforced by the fact that on a clear night people could actually see Sputnik as a star-like point of light moving across the sky. Thanks to sunlight, it reflected back to the night side of the Earth (in fact, what people saw at the time was not the actual Sputnik satellite, but rather the much larger upper stage of the rocket orbiting close by).

Thinking that Russia was a backward country years behind in technology, America had been taken completely by surprise. Totally underestimating Soviet rocket power, some people even thought the comma in the published "83,6 kg" (184 pounds) mass of Sputnik 1 was wrongly placed and that the satellite actually weighed only 8.36 kilograms (18.4 pounds)!

However, the Soviets proved that they could launch even larger objects when only a month later they put Sputnik 2 into orbit. This satellite, weighing an incredible 508 kilograms (1,120 pounds), even carried a dog named Laika on board! Housed in a small, pressured container, Laika served as a test subject for obtaining data on the effects of microgravity on mammals.

As the satellite did not contain a return capsule, the unfortunate animal died seven days later when the oxygen supply ran out. Nevertheless, Laika's flight showed the Soviet's intention to launch people into space; it also showed that this would in principle be possible once a way was found to bring them back to Earth.

Nightmares of Soviet nuclear missile platforms in space and on the Moon made the USA scramble to get its own satellites into orbit. The problem was that the USA had only developed relatively small ballistic missiles for their relatively light nuclear bombs, and these were barely powerful enough to launch anything into orbit.

On December 6, 1957 the United States Navy's fragile Vanguard rocket with its puny Vanguard satellite (Russia's leader Khrushchev called it a "grapefruit") toppled over and exploded on its Cape Canaveral launch pad in front of the television cameras. America's pride was hurt deeply.

Fortunately, von Braun's team had not lost its vision of launching satellites while working on rockets for the US Army. Unofficially and secretly, the German experts had converted a ballistic missile into a satellite launcher. The USA could have had their satellite in orbit before Sputnik, but politics had assigned the space program to the Navy. Now, with the whole nation in a hurry to catch up with the Soviets, von Braun and his Army team were given the go-ahead to launch the Explorer 1 satellite with their Juno 1 rocket.

The launch on January 30, 1958, was a complete success and gave America back some of its confidence in its own technological capabilities. Instrumentation on board Explorer 1 even discovered large bands of charged particles from the Sun trapped by the Earth's magnetic field. These were later named the Van Allen radiation belts, after the American physicist James Van Allen who had designed the particle measurement equipment.

The next big step was to send a person into orbit. With the launch of Yuri Gagarin in 1961 the Soviets were first again, but the Russian cosmonauts were soon followed by NASA astronauts.

As launchers got bigger and satellite equipment technology advanced, Russia and America sent increasingly larger and heavier satellites into orbit. At first their purpose was only to measure the conditions in space, but soon they were also put to practical use such as relaying telephone and television signals and observing the Earth. This has in fact now become the commercially most important application of space.

The Soviet and US governments also soon recognized the military value of satellites, in particular for observing each other's military installations with large telescope cameras and for intercepting radio signals. As each side put ever larger numbers of nuclear missiles into operation and tightened security around them, satellites were the best, and often the only, way to monitor enemy rocket launches and atomic bomb

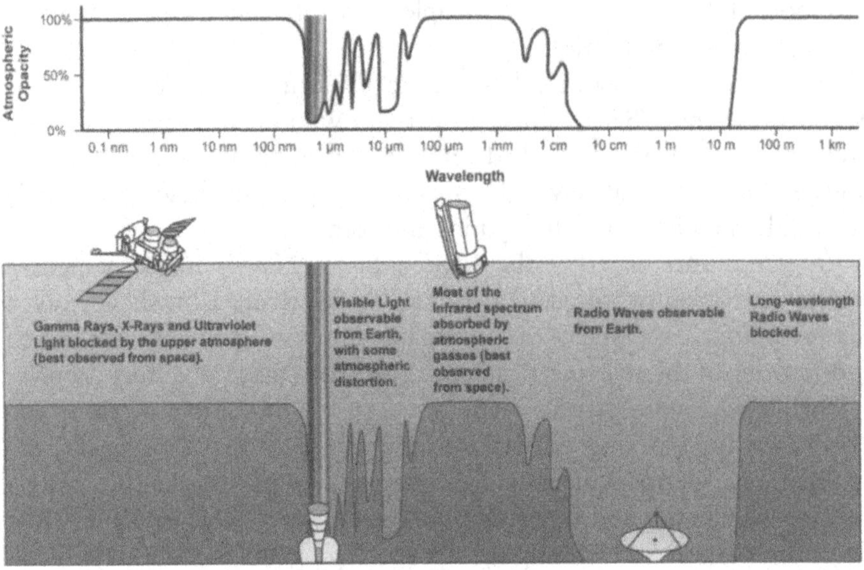

FIGURE 1.5 The atmosphere blocks much of the radiation from the Sun, planets, stars and other astronomical objects; only visible light, some infrared radiation and a large portion of radio waves get through. To see the rest of the spectrum, observations have to be done in space. [NASA]

tests. Increasingly powerful optical, radar and radio-receiving spy satellites are still being launched today, even though the Cold War is over. Many nations still want, and need, to know what their neighbors are building in their back yard.

Astronomy became another important occupation for satellites. High above the Earth's atmosphere – which filters out much of the spectrum of scientifically interesting radiation and disturbs light so that stars appear to twinkle – orbital observatories have an unprecedented clear view on the Universe.

Both the USA and the Soviet Union started to launch ever more sophisticated space observatories, detecting objects in the Universe that were completely unknown until then. The study of exotic objects such as black holes and gamma-ray bursts started entirely new branches of the ancient science of astronomy.

The pictures of the famous Hubble Space Telescope now frequently reach the front pages of the newspapers, but equally interesting data is sent to us by modern X-ray, gamma-ray and infrared observing satellites such as the XMM and Integral of ESA (European Space Agency), and NASA's Chandra Observatory and Spitzer Telescope. They have shown us a Universe that is even weirder than we ever imagined, with all kinds of peculiar star systems, mysterious black holes, gaseous nebulae where new stars are born, and old stars dying in gigantic explosions. However, even though they look at objects millions of light-years away, these satellites still remain relatively close to Earth.

TO THE MOON

From the point of velocity and energy, once you have the ability to reach Earth orbit you don't have to stretch too much to send a space probe further – to the Moon or to the other planets in the Solar System.

To stay in a low Earth orbit at an altitude of 200 kilometers (125 miles), a satellite needs to have a velocity of about 7.8 kilometers (4.8 miles) per second. To reach a much higher orbit at 36,000 kilometers (23,000 miles) requires a boost of only an extra few kilometers per second. This velocity increase is not too hard to achieve, as the difficult part of flying up through the atmosphere with all its problems of aerodynamic drag is already behind when you reach low Earth orbit. Moreover, as you get further from the Earth, the pull of its gravity weakens.

The 36,000-kilometer (23,000-mile) altitude orbit over the equator is a very popular one for television and communication satellites, because there they circle the Earth in 23 hours 56 minutes. (The circle around the Earth they have to cover is much larger than for a low Earth orbit satellite and, furthermore, their orbital velocity is lower; the net effect is that while it takes a low orbit satellite only 1.5 hours to go around once, it takes nearly 24 hours at 36,000 kilometers.) As the Earth also makes one complete rotation in 23 hours 6 minutes, it appears that the satellites are permanently hanging still over the same area, thereby acting as giant, stable antenna masts able to cover about a third of the planet per satellite. This 36,000-kilometer (23,000-mile) circular orbit over the equator is therefore called the "Geostationary Earth Orbit," or "GEO."

To completely escape Earth's gravity and go elsewhere in the Solar System, you'll need to achieve a velocity of 11.2 kilometers (7 miles) per second), starting from the Earth's surface. After achieving that, there is no need to keep the engine running as you would need to do with a vehicle on Earth; in space, once you achieve the right orbit, you cruise effortlessly along through the vacuum, unhindered by any braking friction. You are then in orbit around the Sun, only possibly disturbed by the gravity of other planets.

That is why as early as January 2, 1959, only one year and three months after the launch of Sputnik 1, Russia already managed to send a probe to the Moon. Luna 1 was meant to hit our rocky neighbor, but just missed it and went into orbit around the Sun instead. On September 15, its successor, Luna 2, hit its target spot-on. Crashing into an area called the "Sea of Rain," the metal ball showered the lunar surface with small emblems of the Soviet Union.

Only a month later Luna 3 zoomed behind the Moon, taking the first ever pictures of the part of the lunar surface that is always turned away from us. As the electro-optical systems we have nowadays did not exist at the time, Luna 3 carried a conventional camera with rolls of film. Following exposure, the film had to be developed on board, then scanned by a kind of television camera system. The gray-scales of the scanned picture were coded into radio signals and sent to Earth, where they could be translated back into an image.

Luna 3 showed that there were fewer "lunar seas" (the dark, relatively smooth areas on the Moon we can see with the naked eye) on the farside than on the nearside. As a bonus propaganda caper for the Soviet Union, Luna 3 also broadcast the Socialist International hymn from space. As with Sputnik 1, America was beaten once more.

An even worse propaganda disaster for the USA occurred on April 12, 1961, when Russia managed to launch the first human into orbit. Cosmonaut Yuri Gagarin made one circle around the Earth in 108 minutes and landed safely on the deserted steppe of Kazakhstan. At that time, NASA was still only planning its first sub-orbital, parabolic "hops" with its Mercury capsule and low-power Redstone rocket combination. On May 5, astronaut Alan Shepard became the first American in space, although not in orbit, using this system.

It took almost a year for America to catch up and equal Gagarin's flight; on February 20, 1962, John Glenn rode his Mercury capsule into orbit on top of a more powerful Atlas rocket, and completed three orbits. However, by that time Russian cosmonaut Gherman Titov had already circled the planet no less than 17 times during a flight lasting 25 hours 18 minutes.

Tired of seeing the USA being second in all major space achievements until then, President Kennedy, in his famous speech of May 25, 1961, proclaimed that America should aim at landing a man on the Moon before the end of the decade. Apart from major advances in manned spaceflight, the ensuing "Moon Race" resulted in whole armadas of unmanned lunar orbiters and landers.

The American Ranger spacecraft were first sent on kamikaze missions, aiming straight at points on the lunar surface while taking as many pictures as possible before impacting. Next came the soft-landing Surveyors, meant to determine what the surface of the Moon looked like. Some scientists argued that billions of years of meteorite impacts and unfiltered radiation had pulverized the lunar surface to a depth of possibly many meters; astronauts attempting to land would surely sink into thick layers of dust. Fortunately they were wrong. Digging with their small scoops and making small jumps on their landing gear to be able to observe the depth of their footprints, the Surveyors showed that the dust was only a few centimeters deep at most.

In the meantime, NASA's Lunar Orbiters accurately photographed the surface from lunar orbit to find the most suitable landing spots for the upcoming manned Apollo missions.

The Soviets had already confirmed their unmanned exploration prowess by being the first to softly land a probe on the surface with Luna 9, which made panoramic images of its surroundings. Their next spacecraft, Luna 10, became the first satellite of the Moon.

Nevertheless, the Soviet Union was falling behind in the crewed part of the space race, as unmanned prototypes of its giant N-1 moonrocket consistently blew up shortly after leaving the launch pad. Without a proper heavy launcher, their large lunar capsules and landers were grounded.

The giant Saturn V rocket developed by von Braun and his team was much more successful. After only two test flights, the second of which was only partly successful, the machine was deemed to be ready for operational duty. NASA wanted to use the Saturn V as soon as possible to launch a crewed flight around the Moon before the Russians. The gamble paid off: in 1968 the enormous rocket engines pushed the first Americans in an orbit around the Moon on Apollo 8.

The following year the crew of Apollo 11 thundered off the launch pad. Even as Neil Armstrong, Edwin Aldrin and Michael Collins were heading to the Moon to deliver on Kennedy's promise, Russia desperately sent the uncrewed Luna 15 to land. Its mission was to quickly scoop up some grams of surface samples and return them to Earth before the American astronauts arrived with their load of moonrocks. However, while Apollo 11's crew prepared to leave the Moon and head back home, Luna 15's mission ended with an inglorious crash on the lunar surface. The Soviets had failed to snatch some lunar dust before the American astronauts, leaving them unable to even slightly dim the blazing glory of Apollo 11.

Chapter 1

HUMANS VERSUS ROBOTS

During the "Moon Race" of the 1960s, astronauts and cosmonauts captured the hearts of the public much more than any robotic spacecraft ever could, and proved to be far more flexible in adapting mission operations and handling emergency situations.

Even in comparison with today's advanced robotic Marsrovers, the Apollo lunar astronauts were much more efficient in collecting surface samples and covered far more terrain in less time. The two NASA rovers that were landed on Mars in January 2004 have driven less than 10 kilometers (6 miles) in two years of operation. The record distance for one rover in a Mars day (24 hours 37 minutes) was 220 meters (720 feet). Obviously a person in a spacesuit on Mars could do much better than that.

An astronaut may need a minute to spot an interesting rock, walk to it, grab the stone and inspect it from all sides. For an unmanned rover the same process may take several days.

Moreover, a human may see that the soil under the rock that has just been picked up has a strange color and decide to take a sample of it, while an automatic rover would not detect such a thing unless it was specifically instructed to look for it. A robotic probe can basically only perform tasks it has been built to do before launch, while people are inventive and able to make creative use of what is at hand.

An astronaut may decide to use a spade to break a small piece of a rock from the wall of a canyon a few meters high. Unless a rover has been equipped with a hammer and a long arm, it would not be capable of doing the same.

For a robot, a rock is either an object to avoid or an object to push an instrument against. It is incapable of seeing the rock as a piece of a geological puzzle, or imaging flows of water shaping it into peculiar forms. Computers do not have imagination and creativity, and they are generally incapable of learning by themselves; they literally cannot think "out of the box."

Because they need to be severely modified to survive in space, the computer technology flown in modern space probes is some 10 to 15 years behind what is used in a typical office or at home. With respect to a modern PC, space robots are rather backward. Moreover, even the most modern terrestrial supercomputer is barely able to approach the intelligence of a frog. They may be great in mathematics and even beat us in chess, but reasoning and making decisions in novel situations is not their strong point.

However, the disadvantages of human spaceflight soon showed as well.

To begin with people require a certain minimum amount of living volume, food and oxygen that do not need to be included in robotic spacecraft. People want windows, want to sleep, want to interact with other human beings and have many more physical and psychological demands that are not an issue for robotic space exploration.

People are also continuously "on," whether they are needed or not. During the long flights to their destination, robots can simply shut down or be put in hibernation mode. People need food and oxygen even if there is nothing for them to do and are really not needed for that part of the mission.

Furthermore, people cannot be productive all the time. They need about a third of an Earth day to sleep, and also some time to wash, eat and relax. In principle, robots can work around the clock, day after day, week after week. They do not have holidays and they do not have unions.

Human crewmembers can forget their training and make mistakes. They need to practice just to keep the skills they started with and perhaps improve on them. Robots never forget anything and they do not need to train (although there are now some experiments with "learning" robots that program themselves through experience). Once rightly programmed, robot computer brains are able to refer to the right lines of software at any time. Improvements are possible by uploading updated software, with instant results.

Robots are entirely reproducible; if one fails, you can send an exact copy to take over and you can even skip all the development that was necessary for the first model. In contrast, each person is unique and can never be truly replaced. People cannot simply "download" the training of their colleagues; for each new person you have to start all over again. Robots are expendable; people are not.

Human crews also need to be brought back home after some time. This requires rockets and a lot of propellant, especially when they need to take off from another planet or moon; they require a heatshield for protection against the fierce heat caused by the collision with the atmosphere during re-entry, and parachutes for the final deceleration before landing. Space robots, generally, are simply left at their last destination.

Another disadvantage is that people can die; we have to limit risks for human missions to levels far lower than what would be acceptable for a robotic mission. This involves the duplication of critical equipment, extra design evaluations and tests, and very large ground support teams for monitoring all the systems all the time. For human missions, "failure is not an option," while for unmanned missions risk is just another parameter that can be traded for lower costs and less complex designs.

All this means that even the smallest crewed spacecraft is rather large, heavy, complex and thus expensive compared to an unmanned satellite with a similar amount of scientific equipment on board. To launch heavier and larger crewed spacecraft also requires larger and more expensive rockets.

A typical Space Shuttle mission may cost as much as $500 million, just to send eight astronauts and some experiments in a low orbit around the Earth for less than two weeks. For about the same amount of money you can develop and launch a medium sized robotic orbiter, and investigate the surface of another planet for a couple of years. For about $800 million, less than the price of two Space Shuttle missions, NASA was able to sent two landers, each with a golf cart sized rover, to Mars and operate them for a couple of months. These MER rovers actually lasted well beyond their design lifetime, and for a modest increase in budget they were operated for an even longer period.

The result of all this has been that manned spaceflight has become somewhat stuck in low Earth orbit after the Apollo Moon landings ended. Since the last few astronauts gathered their Moon samples and stepped into their lunar module to return home, we have been operating the Space Shuttle and Soyuz vehicles, as well as several space stations, but gone no higher than a couple of hundred kilometers above the atmosphere. For the last 30 years, flying any further has simply been deemed too difficult, too risky and too costly.

Robotic space probes, however, have long since visited all the eight major planets, flown through comet's tails, landed on asteroids, driven over the surface of Mars and explored the edges of the Solar System. It is a matter of fact that, until now, unmanned spacecraft have taught us far more about the Universe than human missions.

Nevertheless, there are good reasons for sending humans to Mars and beyond, such as the fact that people always want to visit places themselves and that crewed space flights are better than robotic missions for inspiring young people to study sciences. Looking at pictures from Mars taken by unmanned rovers is like having a look at the Grand Canyon through a webcam on the Internet; it is easy, relatively cheap and safe but does not offer the same experience as actually being there.

In the future, some of us may even choose to live on Mars: to increase the survival of our species in case of a global disaster on our mother planet; to expand our civilization; to gain access to the raw materials available on the planets and asteroids; to push our technological capabilities; or just to have the adventure.

In such cases, we should regard our robotic probes as scouts that tell us what lies ahead and can go where we yet fear to tread. They can test technologies that are currently too novel and too risky for integration in crewed spacecraft design.

ESA's Aurora program provides a good example. It is a roadmap that involves a series of increasingly complicated robotic Mars missions, starting with relatively simple demonstration landers and orbiters and evolving to a Mars Sample Return mission that brings Martian soil back to Earth. Each mission will result in more scientific knowledge about the red planet, but moreover will be part of a technology development plan leading to a human mission to Mars sometime around 2035.

Human and robotic space exploration should not be in competition, but rather be regarded as complementary elements of the same push of humanity into space.

FIGURE 1.6 A future astronaut finds her robotic predecessor on Mars. [ESA]

2

A SPACE ROBOT IS BORN

THE term "robot" is derived from a Czech word meaning "forced labor," and got its modern meaning from the 1920 play "R.U.R." (Rossum's Universal Robots) by Czech playwright Karel Čapek.

Simply said, a robot is a mechanism that works according to logical, programmed rules. According to this definition, all spacecraft are robots, and robots have now been exploring the Solar System for nearly half a century.

In Čapek's play, robots develop emotions and overthrow their human masters. Fortunately, for now space robots are doing what we want (save for malfunctions) and emotions are not part of their extensive array of capabilities. If we order them to, they will plunge into certain destruction without a hint of concern or care. They are incapable of being angry or scared, and cannot object, strike or revolt.

Another thing they cannot do is to procreate; leave two of them in the same room and there will still be two many years later. Instead, space robots originate from people's imagination and our continuous yearn for knowledge. We are the robots' creators. We decide on their goals in life and tell them what to do. If they could think, they would consider us to be their gods (actually the theme of the first *Star Trek* movie, in which a vastly enhanced Voyager probe returns to Earth to meet its "Creator").

Space probe mission plans usually start out as a scientific need for information about one of the many fascinating places in the Solar System. There are many targets to choose from: the Sun, the eight major planets, Trans-Neptunian objects, hundreds of small and large moons, asteroids and comets. Scientists may want to study the composition of those beautiful rings of Saturn, or find out whether there really is an ocean of liquid water hiding beneath the thick ice of Jupiter's moon Europa.

All scientific knowledge we gain about one part of the Solar System relates directly to everything else we know or want to learn about our place in the Universe. The crushing carbon dioxide atmosphere of Venus (surface pressure: 90 atmospheres; average surface temperature: 480 degrees Celsius/900 degrees Fahrenheit) shows us what could have happened to our Earth had it evolved a little closer to the Sun. In contrast, Mars shows us what happens if a planet is a little smaller than Earth and receives a little less warmth: the atmosphere has all but escaped the low gravity of the planet and what was once liquid water is now imprisoned as ice below the surface.

The information we ask our robotic science explorers to collect about the Solar System and the Universe is related to the most fundamental questions we want to answer: Why did life begin on Earth? Was it just a matter of chance, or an inevitable result of the planet's favorable position and chemistry? Was there ever life on Mars, maybe when that place was a lot warmer and liquid water freely flowed over its now barren, rusty surface? Did life start anywhere else in the Solar System? And what do the answers to these questions tell us about the chances for other intelligent life in the Universe?

Early wish lists and musings about what questions should be answered, and what data we need to be able to do that, are translated in a set of realistic objectives and requirements for a new space probe mission. Based on these, scientists and engineers together make a preliminary design: what the spacecraft should approximately look like, what it could do and what would be needed in terms of technology, time and money to make it happen.

As you will see, the design of the spacecraft and its mission go through various stages during its development. At each stage the level of detail in the design increases and the new mission comes a bit closer to reality.

Scientific Conception

Nowadays the first spark that starts a new space probe project comes mainly from space scientists. For instance, a group of planetologists (scientists who study planets) may be trying to figure out how the Solar System originally formed.

They think it started as a huge cloud of gas that collapsed under its own gravity. At the center of this contracting cloud the pressure and temperature rose steadily, eventually triggering nuclear fusion reactions that released enormous amounts of energy: the Sun began to shine.

The remaining gas and dust collapsed to a flattened disk around the new, young star. Everywhere in this disk dust particles collided and coalesced, growing in size to form mini-planets, called planetesimals, from which the planets themselves eventually formed.

Closest to the Sun, where temperatures were highest, the four planets made of rocks and metals appeared: Mercury, Venus, Earth and Mars. Farther out, where temperatures were lower, great spheres of gas accumulated around rocky cores (Jupiter and Saturn) or cores made of rock and ice (Uranus and Neptune). In the coldest outer reaches of the Solar System, icy materials collected to form relatively small bodies. These became the Trans-Neptunian objects, including Pluto.

The planetologists have modeled their theories in a computer program that simulates the contraction of the early disk of gas into planets. For some reason however, their computer models cannot accurately predict how the planets closest to the Sun, including our Earth, were born. There is vital information missing about what the inner Solar System looked like some 4.6 billion years ago.

To continue their scientific endeavor they urgently need to know more about Mercury, the planet closest to the Sun. If they knew exactly what it was made of, they could tell what types of material were present near the Sun before the planets were born. If they could figure out whether the interior of the planet is still partly molten or has already completely solidified, they would know something about the time that has elapsed since Mercury was formed.

Unfortunately, not much is known about Mercury. Because it orbits so close to the Sun, the planet is always near its blinding glare and can never be observed in a dark sky from Earth. Pointing the famous Hubble Space Telescope at Mercury is not possible, because direct sunlight accidentally entering the telescope could easily damage the delicate optics and electronics inside.

Sending space probes to visit Mercury is also difficult because it requires a lot of propellant to reach it, and spacecraft going so close to the Sun get very hot. Only one probe, NASA's Mariner 10, has so far visited the little planet. It swiftly flew past Mercury three times in 1974 and 1975, returning the only close-up images of the planet we have. And those pictures cover no more than 45 percent of the surface. Clearly, a new, more sophisticated spacecraft needs to be sent to Mercury to answer the questions that are burning in the minds of our motivated planetologists.

After clearly defining what they need to know about Mercury, the scientists form a first idea about the types of instruments that are required. A set of sophisticated cameras can show them what the surface looks like. Moreover, by counting the number of impact craters on the pictures, the scientists could then roughly determine the age of the surface of Mercury. This would give an indication of when Mercury cooled off and became a solid planet. Unfortunately, cameras alone cannot tell us what Mercury is made of. For that we need instruments called spectrometers, which can analyze the sunlight the planet reflects and thereby give information about the composition of the surface.

A magnetometer instrument can measure Mercury's magnetic field. Earth has a strong magnetic field because it has a solid inner core (consisting mostly of nickel and iron) surrounded by an outer core of liquid metal, and a relatively fast spin rate. The resulting dynamo effect produces powerful magnetic forces. However, Mercury spins rather slowly and we expect its metal core to be entirely solid. It should not have a magnetic field at all, but nevertheless it does, even though it's about 100 times weaker than that of the Earth. Detailed investigations of this mysterious field may result in new theories about what's inside Mercury, and thus about the creation of planets in general.

The choice of instruments leads to questions concerning the spacecraft itself: At what altitude should the probe and thus the instruments orbit Mercury? Stay too high and the cameras cannot see the amount of detail the scientists need. Go too low and the spacecraft moves too fast across the surface, leaving insufficient time to scan areas in detail. In a very low orbit the spacecraft's field of view is also too limited, making it hard to relate the small surface features it sees to the wider context of the planet. Moreover, if it's placed in too low an orbit the spacecraft may crash into Mercury.

A circular orbit ensures that the probe is always at the same altitude above the surface, but takes a lot of propellant to achieve. A highly elliptical orbit is easier to obtain, but then the spacecraft's distance to the planet will continually change. That may be bad for the scientific observations, because the field of view of the surface will constantly differ. On the other hand, close-up observations can be made during the part of the elliptical orbit closest to the planet, while the long time spent further away can be used to radio the fresh images to Earth.

Maybe the scientists decide an orbiting spacecraft alone is not sufficient and they would prefer a lander to actually measure the composition of the soil on the surface. A lander can only make measurements at one spot, so perhaps a series of landers needs to be spread all over the planet. However, too many spacecraft and landers would make the mission too expensive.

In the end the scientists may decide they need to fly a satellite with a multi-wavelength camera system that maps the mineralogy and elemental composition of Mercury, and a magnetometer to measure the strength and structure of its magnetic field. They may also like to have at least one lander to investigate the surface on the spot.

If the lander finds that, in its landing area, silicates are common in the soil, then this can be coupled with the remote-sensing data from the orbiter's camera. Wherever the orbiter's multi-wavelength camera sees the same "colors" as at the lander's site, the soil is bound to contain a lot of silicates as well. In such a way the remote-sensing data from the orbiter's camera can be calibrated using the very local findings of the lander.

After a broad mission plan has been put together it has to be submitted to a government space agency to get funding for the development and operation. In the USA the scientists would try to get an agreement with NASA; in Europe they would contact ESA or a national space agency such as the French CNES, the German DLR or the Italian ASI; in Russia they would go to the Russian Space Agency; and in Japan they would go to JAXA.

Experts in the space agency involved would normally look at the preliminary plan to see if it is worth studying further. Is the envisaged mission important enough to justify spending an important part of the nowadays limited budgets for space? Does it have good chance of success, or is it highly risky or based assumptions that are on too optimistic? Are there sufficient numbers of scientists backing the proposal?

If the verdict is positive, the mission will go into the Conceptual Study phase, and a first rough design of the spacecraft and its mission will be put together.

Considerations such as those described above have actually led to plans for two new, complex space missions called *Messenger* and *BepiColombo*.

Messenger, for *MErcury Surface, Space ENvironment GEochemistry and Ranging*, is a NASA mission that consists of a single orbiter. It will investigate the surface and magnetic field of the planet, and will look for traces of water-ice that may exist deep inside the permanently shadowed craters at the poles. Messenger will map nearly the entire planet in color,

FIGURE 2.1 NASA's Messenger spacecraft being put on top of its Delta II launcher. [NASA]

including most of the areas that could not be seen by Mariner 10. It will also measure the magnetosphere, and the composition of both the surface and the extremely tenuous atmosphere. Messenger should provide us with the first new data from Mercury in more than 30 years.

As Mercury orbits much closer to the Sun than Earth does, Messenger will have to lose velocity to spiral down toward the Sun and reach the planet. Moreover, once there it needs a lot of thrust to get into a nice, relatively low circular orbit. Using an onboard rocket engine for all of this would take a lot of propellant and therefore make the spacecraft very heavy and large. As a result, it would become too costly and also require a very large, powerful and thus expensive launcher.

To stay within budget, Messenger instead uses some peculiar orbital maneuvers called "gravity assists." A gravity assist maneuver means that a spacecraft swings by a planet and uses that planet's gravitational field and orbital velocity around the Sun to pick up or lose speed. In essence, the spacecraft uses some of the planet's energy for its own speed changes. Since the planets are much, much more massive than a spacecraft and thus have vastly more orbital energy, this parasitic behavior has no measurable effect on the planet's orbit. However, for the spacecraft the "stolen" energy can be used for large velocity and orbit direction changes.

In case the energy is used for increasing speed, the situation can be somewhat compared to a ping-pong ball hitting a revolving fan, which will bounce back at a much higher speed than the speed at which you threw it. Another analogy is that of a car picking up speed by free-wheeling down a steeply inclined road, but instead of the Earth's gravity, the probe uses the gravity of another planet. Messenger will use the principle in reverse to lose enough velocity so that its new orbit matches that of Mercury around the Sun. It's like a car slowing down by driving up a steeply inclined road.

After launch, Messenger will circle the Sun and return to Earth for its first gravity assist maneuver in August 2005, then fly past Venus in October 2006 and again in June 2007 (see Figure 2.2). These smart, propellant-saving maneuvers come at a price, however; although the probe has been launched in August 2004, it will not arrive at Mercury before 2011. The multiple sweeps past Earth and Venus consume a lot of time.

BepiColombo is ESA's mission to Mercury and in its current design consists of two separate orbiters that are planned to be launched in 2012. One spacecraft will be dedicated to imaging the planet, while the other will mainly study Mercury's magnetic field from another orbit. Initially, there were plans to include a lander as well, but that was found to make the mission overly complex and costly (see Figure 2.3).

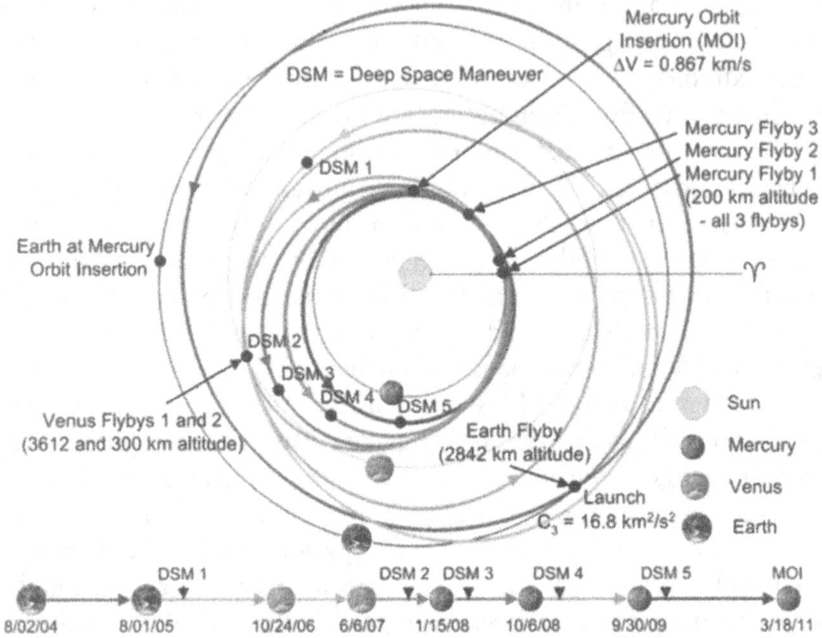

FIGURE 2.2 The complex trajectory of Messenger on its way to Mercury. Various planetary flybys and Deep Space Maneuvers with its rocket engine will bring it into the right orbit. [NASA]

The project is named after professor Giuseppe (Bepi) Colombo, who lived from 1920 to 1984 and dedicated his life to detailed studies of Mercury and interplanetary spacecraft trajectories. His authority was widely recognized.

When in 1970 Guiseppe was invited to participate in a NASA conference on the planned Mariner 10 mission, he realized that after flying by Mercury, the period of the spacecraft's orbit would be very close to twice the orbital period of the planet. In the time it took the spacecraft to make one orbit around the Sun, Mercury would have gone around twice. He suggested that, if the initial fly-past point could be carefully selected, the two could meet each other more than once.

NASA adopted his brilliant idea, and eventually Mariner 10 made three flybys of Mercury in 1974 and 1975. Thanks to Guiseppe Colombo's orbital calculations, the spacecraft was able to collect far more scientific data than would have been possible during a single encounter.

For ESA the BepiColombo mission is very special, as most of its previous interplanetary missions have gone outward, to the relatively cold parts of the Solar System. Instead, BepiColombo will be flying into the really "hot" inner regions where the Sun's radiation threatens to fry a

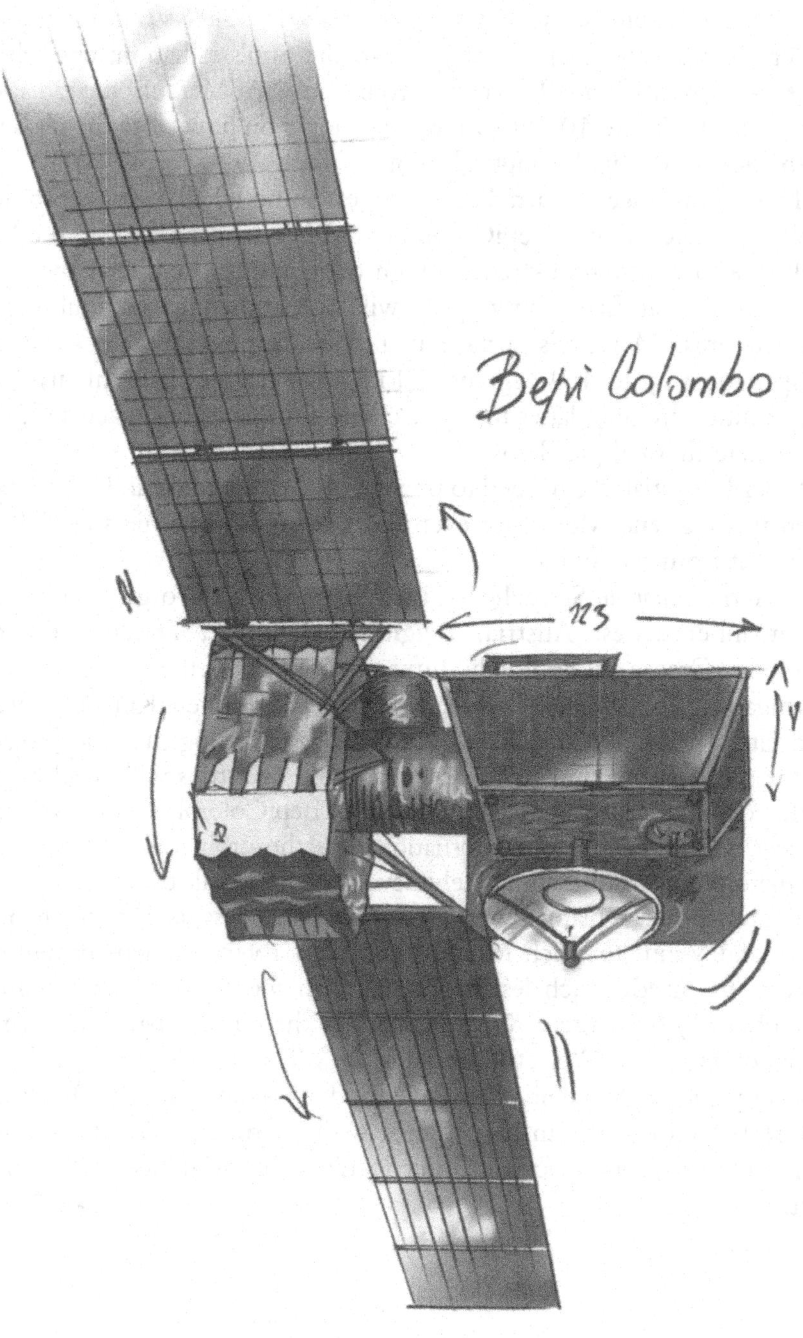

FIGURE 2.3 An early sketch of ESA's BepiColombo spacecraft [ESA]

spacecraft. ESA's only other spacecraft that was sent that way was Venus Express, but BepiColombo will get much closer to the Sun.

A satellite orbiting the Earth receives about 1,400 watts per square meter (10.5 square feet) of energy from the Sun's radiation; enough to light 14 lamps, each of 100 watts, in your house. At Mercury the radiation energy can be five to 10 times more, depending on how close the planet is to the Sun in its highly elliptical orbit.

This abundance of irradiation poses some challenging technical problems. The sides of BepiColombo and Messenger that face the Sun will have to withstand extremely high temperatures, while, at the same time, those that face empty space will be exposed to incredibly low temperatures. Mercury's surface is also heated by the Sun, reaching temperatures up to 450 degrees Celsius (840 degrees Fahrenheit). The heat emitted by the planet into space, and thus onto the spacecraft, can also create thermal problems.

Special materials are needed to prevent outside antennas and solar arrays from melting, and Messenger even hides behind a ceramic sunshade to protect it from the Sun.

ESA, the *European Space Agency*, is a very international organization with 17 member states: Austria, Belgium, Denmark, Finland, France, Germany, Greece, Ireland, Italy, Luxembourg, the Netherlands, Norway, Portugal, Spain, Sweden, Switzerland and the United Kingdom, with associated members Canada and Hungary participating in some projects under cooperation agreements. Each of these countries will develop and build equipment for the imaging orbiter of BepiColombo. However, this project will be even more international, as the orbiter for studying the magnetosphere will be completely developed and built by Japan.

BepiColombo is planned to use not only gravity assists from the Moon, Earth and Venus to reach Mercury, but also a very efficient propulsion system that needs much less propellant than the conventional chemical propulsion systems used by Messenger. In the current baseline designs, BepiColombo uses "electric propulsion." The use of this for interplanetary flights is rather innovative, and therefore needs to be tested before BepiColombo can be designed with confidence. Electric propulsion and the special spacecraft built to test it is the subject of the next section.

PROVING TECHNOLOGY

Most modern interplanetary space missions have a primarily scientific motivation. However, sometimes it is technology, not science, that is the main goal of the mission.

ESA's SMART 1 is such a project. Although the SMART 1 probe flew to the Moon, the main requirement for the mission is to test and demonstrate new technology.

SMART stands for "Small Missions for Advanced Research in Technology," and SMART 1 is the first in what should become a series of technology demonstration satellites. The 370-kilogram (820-pound) SMART 1 carries a suite of miniaturized scientific instruments that are making an inventory of the key chemical components of the lunar surface.

Even after the explorations by the Apollo astronauts on the Moon in the 1960s and 1970s, there is still a lot of vital information missing about the Moon's composition. Without this, we cannot determine how our neighbor in space has been created. Is the Moon formed from debris of a violent collision of a small planet with the young Earth? Or has it formed

FIGURE 2.4 ESA's SMART 1 spacecraft is testing the use of electric propulsion for interplanetary flight. [ESA/J. Huart]

somewhere else in the Solar System and later been captured by the gravity of our planet, imprisoning the Moon in a tight orbit around the Earth?

The main innovation of the SMART 1 mission is, however, its revolutionary propulsion system. Where most interplanetary probes rely on propellant-hungry chemical rocket engines, SMART 1 is testing a much more efficient electric propulsion system called an "ion engine." The technology was pioneered by NASA's Deep Space 1 mission, but had not yet been thoroughly developed and tested in Europe.

Ion engines make use of an effect that was discovered by the American physicist E.H. Hall in 1879. He observed that when an electric current (through a wire) flows across a magnetic field, it creates an electric field

FIGURE 2.5 The ion propulsion thruster of ESA's SMART 1 spacecraft. Ions are ejected at high speed, resulting in a small but very efficient means of thrust. [ESA]

directed sideways to the current. When electrically charged atoms, called ions, are placed in this electric field, they can be accelerated to great speeds.

The ion engine of SMART 1 uses the inert gas Xenon as propellant. The atoms of this gas are ionized and then shot out in a constant flow. Owing to the "action = reaction" principle explained in Chapter 1 of this book, the Xenon ions shooting out of the engine push the spacecraft forward.

In an ion engine, the gas is ejected at a much higher velocity than in a normal chemical rocket: an engine burning liquid propellants may shoot out hot gasses at 4.4 kilometers (2.7 miles) per second, but SMART 1's ion engine expels its charged atoms at more than four times this speed. Because of this higher ejection velocity, ion engines are much more efficient and thus need much less propellant. The propellant saved means that more space and mass are available for scientific instruments; it also means that interplanetary spacecraft can be made smaller and lighter than before.

On SMART 1 the energy needed for the ion engine comes from its solar panels, which convert sunlight into electricity. The technique employed by SMART 1 is therefore officially called "solar-electric propulsion."

SMART 1's ion engine is very efficient. Flying to the Moon, the little spacecraft traveled 84 million kilometers (52 million miles) with a fuel consumption of less than 2 million kilometers per liter (4.7 million miles per gallon).

However, there is a catch when using solar-electric propulsion: ion engines are amazingly propellant-efficient but only give a very low thrust. It's like driving a car with an extremely economical but not very powerful engine, going 500 kilometers (310 miles) on one liter of gasoline but at a maximum speed of only 30 kilometers (18 miles) per hour. SMART 1's engine pushes the spacecraft with a force of only 7 grams – less than the weight of a postcard. This extremely low thrust results in a very low acceleration of the spacecraft; each second, with the engine on, SMART 1 increased its velocity by only 0.2 millimeter (0.01 inch) per second.

With a couple of short boosts from a conventional rocket engine you can reach the Moon in a couple of days, but it took SMART 1 16 months to get there. The gentle push meant that SMART 1's ion engine had to work for the entire duration of the transfer, slowly increasing the diameter of its orbit around the Earth until it reached the Moon. One of the most important tests of the mission was to see whether the engine could really work continuously for over 16 months. SMART 1 had to fly 332 times around the Earth before finally attaining lunar orbit.

If the study of the Moon had been the primary goal, it could have been done faster and cheaper with a simple spacecraft based on a conventional chemical rocket system. The requirement to test electric propulsion makes SMART 1 a very unconventional lunar mission.

Ultimately the reason for developing and testing new technology on board SMART 1 is to make future space missions with electric propulsion possible with lower risks and at lower cost. What we learn from SMART 1 can be applied to larger, purely scientific missions such as BepiColombo, which will go much further than the Moon. Putting experimental, untested technology on board an expensive mission such as this would be too risky, but through SMART missions the new technology is mastered and becomes "flight proven." SMART 1 is thus not a stand-alone mission, but part of a grand, comprehensive plan of future interplanetary exploration.

The Political Push

Sometimes it is not only the pure scientific or technical issues that define the requirements. During the space race of the 1950s and 1960s, politicians and military planners had a strong influence on which missions got funding and what the spacecraft needed to do. At that time, prestige and propaganda were very important – and often more important than the scientific benefits.

If your country was the first to land something on the Moon, you would make the international newspaper headlines and gain the admiration of the world, even if your probe hardly carried any scientific instruments. In the early stages of spaceflight, merely hitting the Moon was a major achievement, and one that could be exploited politically. The Soviet Union and America both tried to show the rest of the world that their social and economic systems were the best. One way of proving that was by superior technical achievements, such as landing on the Moon or flying past Mars.

Moreover, if your rockets were able to hit the Moon, that demonstrated that you were certainly also capable of hitting your enemy's territory on Earth with nuclear missiles. Space missions could be used to showcase Cold War military power without actually hurting anyone.

Nowadays the political push behind space missions is much less strong and a lot more benign. Complicated space projects are a great way to strengthen the ties with other nations through international collaboration.

FIGURE 2.6 Space missions have propaganda value, as shown by this Cuban stamp with an image of the Venus lander Venera 10.

Many American interplanetary spacecraft carry scientific instruments from Europe, and vice versa. NASA's Cassini spacecraft carried ESA's Huygens probe to Saturn and released it to investigate the mysterious atmosphere of its largest moon, Titan, and ESA's BepiColombo mission to Mercury will include a Japanese orbiter.

Apart from using NASA's own Mars Global Surveyor and Mars Odyssey orbiters, the NASA Marsrovers "Spirit" and "Opportunity" could communicate with Earth via ESA's Mars Express orbiter. The British Marslander "Beagle 2," carried on board Mars Express, would have been able to communicate via its European mother ship as well as both NASA orbiters (unfortunately something went wrong on the way down to the Martian surface, and Beagle 2 was lost).

Moreover, ESA itself is a group of European member states that together finance, develop, launch and operate spacecraft. All its projects are highly cooperative and international undertakings.

Spacefaring nations also help each other by offering their radio ground stations to communicate with other nations' space probes. No country is so large that its receivers can always be in the right position to pick up the signals from its own interplanetary spacecraft; because of the rotation of the Earth it will have to depend on stations in other countries to relay communications during "black-out" periods.

Yet another way of cooperating is to offer a free launch on one of your rockets for someone else's spacecraft – in return, of course, for access to the scientific results of the mission. As there is no such thing as a "free lunch" in commercial business, nor is there a "free launch" in space business.

Chapter 2

PAPER SPACECRAFT

Once a general idea for an interplanetary space mission has been formulated and accepted, the real hard work begins. The initial idea, which may be nothing more than a couple of pages with scientific objectives and requirements, needs to be worked out in detail.

First, a list of technical requirements is derived from the scientific requirements. For instance, the scientists may want to have a robotic rover driving over the surface of Mars to investigate the rocks there. A technical requirement following from this is that a system needs to be developed to land the rover softly on Mars.

Moreover, the fact that the rover will be on Mars results in technical requirements about the minimum and maximum temperatures the rover must be able to survive while on the surface. From previous Marslanders and orbiters we now know that the coldest temperature on the surface is about –140 degrees Celsius (–220 degrees Fahrenheit) at the polar caps, while it gets no warmer than 20 degrees Celsius (68 degrees Fahrenheit) anywhere else. As this is too cold for the electronic equipment on board a rover, a heating system is required.

Sometimes the technical requirements cannot be captured into such sharply defined parameters. For instance, when ESA's Huygens probe was designed to investigate the murky atmosphere of Saturn's moon Titan, little was known about the temperatures and pressures a spacecraft would encounter. The probe had therefore to be designed to operate in a relatively wide range of possible atmospheric environments – which was a bit like designing an all-terrain car that has to be able to drive well on both asphalt roads and muddy sand paths.

Moreover, if the probe would make it all the way to the surface alive, would it then hit a super-cold icy surface as hard as steel, or plunge into a bitterly cold lake of liquid hydrocarbon gas? Although Huygens was not specifically intended as a lander, scientific measurements from the surface would provide a valuable bonus to the mission. Huygens, therefore, should not only be able to land on hard, solid ground but should also be able to float, and in both cases be able to gather data.

Starting from the scientific and technical requirements, the spacecraft's mass, the amount of electrical power needed to run all the scientific and supporting equipment, and the size of the probe needed to fit all the hardware and propellant have to be calculated. Moreover, experts need to precisely list the required types and numbers of sensors, solar cells, antennas, tanks, computers, etc.

Typically the main issues the scientists and engineers need to tackle are how to make the probe as light and as small as possible, while, at the same time, achieving all the mission objectives. Robotic interplanetary spacecraft typically have a launch mass somewhere between 300 and 3,000 kilograms (660 and 6,600 pounds), while their size may vary from that of a washing machine to the dimensions of a large shipping container.

In general, the smaller the spacecraft the less costly it is to develop and build. Moreover, the lower the mass and volume, the smaller and thus cheaper the rocket required to launch it. For each kilogram/pound you may save on your interplanetary space probe, you may save some 60 kilograms/pounds of propellant and 7 kilograms/pounds of launcher hardware.

If you save enough on your spacecraft mass, this means you can switch to a smaller and cheaper launcher. As the launch price typically comprises 30 percent or more of the total mission cost, such savings are very important. Making your space probe small and light enough to fit inside a smaller rocket instead of a larger one easily saves you several tens of millions of dollars.

Detailed questions to be answered during the spacecraft design work are, for instance:

- How many propellant tanks should we fit, and what volume should they hold to carry all the propellant?
- How many thrusters (small rocket motors) are enough to stabilize the spacecraft?
- Where do we place the antennas so that at least one is positioned just right at any time to exchange radio signals with Earth?
- How much surface area do we require for the solar arrays to make sure that, even if the efficiency of the solar cells degrades over time, we always have sufficient power to run all the onboard systems?

Many times the scientists and engineers have to cope with conflicting requirements. For instance, to remain cool a spacecraft orbiting Mercury should stay in the planet's shadow as much as possible. If it didn't, its electronics would get fried very quickly. On the other hand, the spacecraft also needs sufficient sunlight to get enough electrical power from its solar arrays. The engineer designing the thermal control subsystem of the spacecraft wants to hide it behind Mercury all the time, while the engineer responsible for the electrical power supply would like it to be exposed to full sunlight as much as possible.

Keeping the spacecraft in the shadow at all times means that it will receive no energy and will die when the battery eventually empties. Facing the Sun continuously requires very complex thermal protection with thick layers of insulation and large radiators. This would make the probe too heavy and too large. Clearly, a compromise has to be found between the preferences of the thermal control engineer and the electrical power engineer.

In addition, the scientists may have very strong requirements for the amount and direction of light they need for their cameras, and that must also be taken into account.

The designing of space probes is not a garage project in which you just scribble a rough sketch on a napkin and start building almost immediately, working out the details as you go. This process only works in some game shows, where contenders have to quickly build machines from whatever they can find in the scrap yard, making adjustments as they progress and fixing problems or even modifying their design during the contest.

Spacecraft are much too complicated for this. Let's say that you have started to build your robotic Jupiter explorer and later find that your battery is too small. You will have start again and incorporate a bigger one, which is of course also heavier. As a result, your spacecraft increases in mass, and therefore needs not only more propellant for its orbit maneuvers but also larger tanks. This makes the whole thing even heavier and larger, requiring the spacecraft's basic structure to gain weight. Also, you will need more thermal isolation material to cover the now larger spacecraft. Such snowball effects can quickly get out of hand.

Furthermore, unlike most other machines, interplanetary probes cannot be repaired after launch. Broken equipment cannot be replaced, so spare units have to be actually built into the spacecraft. This is called "built-in redundancy." If something breaks along the way, such a backup system may take over from the failing equipment, but otherwise there is not much you can do about it. The spacecraft is far out, and gone forever. Your only link with it is via radio communication.

Spacecraft therefore need to be carefully designed on paper (or in computers actually) before anything can be built, bought or constructed. Mistakes made early in the project can come back to haunt you years later, during the detailed development, construction, testing or even when the spacecraft is already flying through space.

Designing spacecraft is not a linear process where you start with some requirements and then simply follow a list of things to calculate and decide on. Instead, space probe engineers work through so-called iterative "design loops."

For instance, when you want to land something on Mars you will normally use parachutes. However, since the atmosphere of Mars is very thin, these may not brake your lander enough to prevent it from smashing into the rocky surface. You will need something else to slow the spacecraft down just before it reaches the ground.

There are two basic possibilities. One is to use a system of braking rocket thrusters to slow the lander down and softly place it in the red Martian dust in a fully controlled way. The benefits of this system are that you can land really carefully and that you always touch down with the right side up. The downside is that you will need a sophisticated control system, instrumentation to very precisely measure the altitude, and a lot of

FIGURE 2.7 The Viking lander used a heatshield, parachutes and rocket engines to land softly on the surface of Mars. [NASA]

rocket propellant. The two Viking landers that NASA sent to Mars in the mid-1970s were based on this system.

The other method is to enclose your spacecraft inside a cluster of airbags (similar to those used in cars) that inflate immediately before hitting the surface, and just let your lander fall, bump and roll until it stops. This results in a much simpler, smaller and lighter system than the use of braking rockets. However, even with the baggy protection the landing is relatively hard; the lander has to endure something that can be compared to pushing your computer from a chair onto the floor here on Earth. Because you lander has been bouncing over the surface, it is shaken quite a bit, and if your spacecraft is heavy the airbags have to be enormous. Moreover, you can never be sure which side is up when the thing finally stops, so the lander has be able to right itself mechanically if it is upside down.

The famous 1997 NASA Mars Pathfinder used this system for the first time, with success. The lander hit the Martian surface with a total velocity of about 65 kilometers (40 miles) per hour; approximately 45 kilometers (28 miles) per hour in the vertical and 45 kilometers (28 miles) per hour in the horizontal direction. It then bounced some 15 meters (50 feet) into the air, bounced another 15 times and rolled over and over before finally coming to rest about 1 kilometer (0.6 mile) away from the initial impact

FIGURE 2.8 Airbags are used to protect a lander when it hits the surface of a planet. The landing is however much rougher than with a rocket brake system [NASA]

site. It had been rolling and bouncing for some 2.5 minutes. More recently, the two NASA Mars Exploration Rovers Spirit and Opportunity were safely landed with airbags.

To make a trade-off on which of these possibilities to use, you could make two preliminary designs: one with a rocket system like Viking and one with airbags like Pathfinder. You could then evaluate the complexity and estimate the masses, sizes and costs for each option and compare them with the original list of requirements for the maximum launch mass, maximum cost and maximum landing speed. The one that best fits the requirements is the winner of the little Darwinian competition.

However, often the trade-offs are not as straightforward as in this example. In his *Akin's Laws of Spacecraft Design*, professor David Akin of the University of Maryland comments: "There is never a single right solution. There are always multiple wrong ones, though."

After all major trade-offs have been made, you can design the chosen concept in more detail. When you have checked the outcomes with the requirements for maximum mass, maximum landing speed, etc., you have finished the first design loop of your new space project. If the concept does not entirely fit the requirements, you will have to do another design loop involving changing your design, recalculating everything and comparing the new numbers with the requirements. If you did it right, your updated design will fit the requirements better than the earlier concept.

Sometimes there is no way to nicely adhere to all the requirements. Your concept may fit the mass and volume demands but be much too expensive. Or it may be too complex to build within the specified timeframe. It has then to be determined whether the requirements themselves need to be changed, or whether the project must be branded "not feasible."

No design is complete without a development schedule plan that shows when to start the development and production of a part of the spacecraft, and when that part needs to be finished. It's no use testing the spacecraft's telecommunication subsystem if the transmitter is not ready yet; it would be like trying to test drive a car while the engine is still being assembled in the garage. Just like an orchestra, where each musician has to be at the same page of the music book and every instrument has to start at the right time, the development and production phases of all parts of a spacecraft need to be harmonized.

The entire development of an interplanetary robotic probe takes about three years minimum, but large and complex spacecraft may take as long as eight years to be ready for launch. These long development times mean that when they are launched, space probes often contain equipment that is already obsolete. When you take into account the projected very long travel times, spacecraft are often fairly antiquated by the time they reach their destination.

When the Huygens probe entered the atmosphere of Saturn's moon, Titan, in January 2005, it had been on its way for over seven years, while its development actually started in the late 1980s. That meant that the technology descending through that strange Titan atmosphere actually dated from the time we were running Microsoft Windows 2.0 on 386 computers!

While designing the spacecraft for minimum mass, volume and complexity, the risk of failures always needs to be taken into account. Is one onboard computer enough? What if it breaks? Maybe you need to put in a backup, but would that make the probe too heavy?

What is the chance that one of the two foldable solar arrays does not deploy properly? Do you design the spacecraft so that it can operate with only one solar array, and use the second one only as a backup? Or do the scientists accept that if one solar array does not work, they will need to operate the scientific instruments only part of the time and thus obtain less valuable information?

The level of risk is evaluated by looking at the combination of impact and likeliness of something going wrong. The impact of aliens hi-jacking your space probe on its way to Mars would, for instance, be very high, because if that happens the mission would be completely lost. However, the probability of such an abduction is extremely small. The risk, as function of the impact and likeliness, is negligible because the high impact is nearly nullified by the extremely low probability of occurrence.

In contrast, the chance that a piece of electric wire will get damaged during the construction of the spacecraft is rather high, but the impact is very low. You just keep some inexpensive spare wire at hand to replace it.

Risks with a relatively high impact and likeliness need to be dealt with by, for example, having some reserve money included in the development budget to buy replacements for faulty equipment.

Relatively high risks connected to equipment failure during the mission can be diminished by the earlier mentioned built-in redundancy – that is, having spares on board that can take over when needed. For instance, two radio transponders (combined transmitters and receivers) are usually included, because without the means of communication any spacecraft is rendered useless. Since this is now more or less standard practice, the two transponders often come integrated as one single box (this is called "internal redundancy").

Another way to lessen the risk of losing your mission is to design so-called functional redundancy into the spacecraft, by ensuring that other types of equipment can take over the job of failing devices. For instance, when sensors that orient a spacecraft by pointing it at a certain star (so-called star trackers, see Chapter 3) break down, their function may be taken over by other types of sensors that aim the probe with the help of the Sun.

There are also risks that are relatively high, but cannot really be dealt with. For instance, the chance that your space probe is lost due to a launch failure is not negligible, even after more than half a century of experience in launches. The impact would be extremely high, because when it happens the spacecraft will surely be destroyed. However, there is not much you can do about it; building a copy of your spacecraft to keep as a spare is much too expensive. It is a risk that cannot be neglected but just has to be accepted. The only thing you could do is to financially insure the project against a failed launch.

Increasingly important in any space project is the planning of a solid financial budget. Has sufficient money been allocated to do the whole mission? What are the risks of over-expenditure? Estimates of the costs for development, production, launch and operations of the probe have to be made, and have to be compared with the actual budget available.

An average space probe, mostly based on existing and somewhat modified equipment, may cost something in the order of 100 million dollars for the basic equipment and another 50 million for the scientific instruments on board. The launch with a Soyuz rocket would cost at least 40 million dollars, and another 20 million dollars would be needed to track and control the spacecraft during flight. That's a total of 210 million dollars. A complex mission, requiring lots of new developments and carrying an extensive array of sophisticated instruments, can easily cost over a billion dollars (see Table 2.1).

Spacecraft are usually unique; they need to be specifically designed for a certain mission and therefore normally only one is built of any type. Even equipment that can be used on more than one space probe, such as standardized solar arrays or antennas, is only built in small numbers. As a result, space probes do not benefit from the cost-reducing economies of mass production, such as we see with computers and cars.

Moreover, the fact that no one can repair interplanetary spacecraft once they are launched, and that they have to operate for long periods in the hostility of space, means that only components of very high quality can be used. That may, for instance, mean that special electronic components need to be used that are especially resistant to disturbances by the higher radiation levels in space. Sometimes several tens of units of the same component need to be bought and tested, so that the best single unit can

TABLE 2.1 Comparisons of the costs for interplanetary space missions and other items

	Cost [$ x 1,000]	**So with one billion $ you can . . .**
McLaren F1 LM sport car	1,300	buy 770 McLaren F1 LM sport cars
Ferrari Enzo sport car	640	buy 1,560 Ferrari Enzo sport cars
F16 fighter plane	22,000	buy 46 F16 fighter planes
Boeing 747 airliner	230,000	buy 4 Boeing 747 airliners
B-2 Spirit stealth bomber	2,200,000	buy half of a B-2 Spirit
Virginia class submarine	2,600,000	buy 40 percent of a Virginia class submarine
Space Shuttle mission	500,000	launch two Space Shuttle missions
Soyuz space tourist trip to the International Space Station	20,000	make 50 trips to the International Space Station as space tourist on board a Russian Soyuz spacecraft
NASA/ESA Cassini–Huygens mission	3,400,000	pay 30 percent of the entire NASA/ ESA Cassini–Huygens mission to Jupiter
Development, launch and operation of the NASA Mars Excursion Rovers	804,000	develop, launch and operate two identical NASA MER-type rovers
Development, launch and operation of ESA's Mars Express orbiter	250,000	develop, launch and operate four different ESA Mars Express-type orbiters
Development, launch and operation of NASA's Lunar Prospector	65,000	develop, launch and operate 15 different NASA Lunar Prospector-type orbiters
A bottle of beer or lemonade	0.001	buy every US citizen four bottles of beer or lemonade

be identified. Only that one unit selected from the whole group may then be used in the actual spacecraft.

However, sometimes costs can be reduced by building spacecraft out of left-over spare equipment and test models of other missions. When ESA's four Cluster spacecraft were all lost during the failed launch of the first Ariane 5 rocket, one nearly complete spacecraft test model and some spare equipment were used in building the replacement satellites of Cluster II. NASA's Magellan Venus orbiter even incorporated all kinds of equipment left over from several previous projects that had totally different mission objectives.

Space projects are notorious for their initial underestimated costs (often done to insure that funding for new missions would be approved by politicians) and consequent overspending at a later stage. Few spacecraft ever get finished under budget; most need additional funding as the development progresses and the project is found to be increasingly expensive.

However, more and more space probes are "built to cost," meaning that the maximum financial budget is set before the designing of the spacecraft is started. If the cost estimates for the first design indicate that it would probably cost more to develop than is allowed, the probe needs to be redesigned until it fits the budget. Often it means that the scientific return of the mission is lowered, for instance, by deleting instruments from the spacecraft to make it lighter, simpler, smaller and thus cheaper.

There is an important difference in how projects are funded between NASA and ESA. NASA projects have to battle for money through the federal budget each fiscal year. This means that even though a project may have been getting sufficient funding for half a decade, there is no guarantee that there will be money to complete it in the following year.

For ESA missions, the budget for the whole project is more or less secure once it has been given the go-ahead; the agency does not have to convince politicians of the need for further funding every year. However, if the initially allocated budget is found to be insufficient, political approval for additional money has to be obtained, of course. Most space scientists and project managers believe that ESA's way is better, at least for their nerves.

Compared to how robotic explorers uniquely enrich our knowledge and understanding of the Universe, the financial investments required are actually quite small. The missions of the two Voyager spacecraft that showed Jupiter, Saturn, Uranus and Neptune and its moons in unprecedented detail have cost each US taxpayer less than a simple lunch. And for the cost of a couple of beers per year the European citizens have been able to explore the Sun, the Moon, Venus, Mars, Saturn's moon Titan and Halley's comet.

Today, NASA's total budget represents less than 1 percent of the total federal expenditures. Even at its peak, in 1966, the space agency's budget was only 5.5 percent of all the money spent by the US government. In Europe the ESA member states spend even less than the USA on space exploration.

In the end, the design reports, blue prints and computer simulations must together show a design that is technically coherent, able to fulfill its scientific objectives with a reasonable amount of certainty, and can be developed and built in a reasonable time and for a reasonable budget.

Chapter 2

IT'S JUST A PHASE I'M GOING THROUGH

In NASA and ESA, spacecraft projects go through a number of strictly defined phases during their development.

The "Pre-Phase A" or "Feasibility" stage is the time during which a conceptual study is done. The spacecraft's general layout is determined, its mass and volume are calculated, the possible interplanetary trajectories are shown and the expected scientific return and chance of success are evaluated.

If the design meets the expectations, it may be approved to move to "Phase A." During this preliminary analysis phase the design is specified in further detail: the exact trajectory the spacecraft will take; when it needs to be launched and when it will arrive at its destination; the tests that need to be performed; the companies that will supply particular equipment; and many more details. Phase A ends with a preliminary plan on what the spacecraft will look like and how to develop, build, launch and operate it.

In "Phase B" the design of the spacecraft is defined in considerable detail, down to the lowest level; requirements and schedules are defined for each sensor, thruster, antenna, mechanism, electric heater, etc. Most importantly, during Phase B the scientific instruments that will be flown on board the spacecraft are decided and the scientific or technical institutes (e.g. university departments, government laboratories, etc.) that will be responsible for them are determined.

Space companies show their Phase B designs to the responsible space agency, and offer formal proposals for further work on the project. Usually several large companies propose different designs for the same project as Prime Contractor – that is, the company that will be responsible for the overall development leadership. Each of the competitors evaluates and proposes the subcontractors it will need to help to develop and supply the various equipment. In the end, the company with the overall best offer wins and gets selected by the space agency to proceed to the next phase of development as Prime Contractor.

Even although their designs may be based on the heritage of very different family lines of previously built spacecraft, probes designed for the same purpose by different companies often look very similar. That is because, out of the vast number of possible spacecraft configurations, usually only a handful of concepts hold the promise of achieving the mission objectives in an optimum fashion.

It therefore often happens that space probe concepts based on the same requirements, but designed independently by company A and company B, look strikingly similar. It is not because they have been spying on each other; it is because both are bound by the same laws of physics and have followed similar engineering logics.

The greatest engineer of all, Mother Nature, shows the same consistency. Ichthyosaurs and dolphins are both animals optimized for living in the sea and chasing fish. Both have streamlined bodies and flip their powerful tails to achieve great speed. Neither is very big and thus agile enough to go after small, fast prey. Ichthyosaurs and dolphins share a very common way of living and therefore look very similar. However, dolphins are modern mammals, while ichthyosaurs were reptiles that have not been around for the last 90 million years. Same requirements and same configuration – in biology, this is called "convergence."

Phase B is followed by "Phase C/D," the combined design and development phase during which the spacecraft concept is detailed down to the last nut and bolt, and the actual building of the spacecraft starts. Equipment like sensors, antennas, tanks and cables are put together to form subsystems for attitude and orbit control, propulsion, electrical power supply, data handling (computers), telecommunication, structures and thermal control. Once integrated with each other, these subsystems form a complete spacecraft.

Usually equipment is not only built for the actual spacecraft, but also for various tests that are necessary during development. For instance, a fuel tank qualification model may be used for pressure tests. During such tests the pressure inside a tank is increased up to a level that is much higher than anything it would normally will have to cope with. The qualification tank may be tested to destruction to discover the exact limits of the design. The design of the actual flight model of the tank to be put on board the spacecraft can then still be improved if necessary.

The equipment may then need to be tested with other equipment to find out how they operate together. A fuel tank model may, for instance, be integrated with test rocket thrusters and the necessary pipes, valves and propellant filters to evaluate the proper functioning of the whole propulsion system.

Finally, all the equipment is assembled into a complete (test) spacecraft, which has to be tested as a whole. This is the topic of Chapter 4, "Building and Testing."

Other tests may be necessary for other types of equipment. The more complex, new and challenging the design, the more tests and thus the more development and test models are needed.

After Phase C/D is completed, the spacecraft is ready to be launched (Phase E), and start Phase F, its "Operations Phase." In addition, there is often a Phase G, the decommissioning phase that deals with the spacecraft's destination at the end of its life. Sometimes spacecraft are just left to continue their journey through space as inert pieces of metal, or as abandoned relics on the surface of a planet. However, some missions don't just fade out because of empty batteries or lack of budget to continue operation, but have a much more exciting ending. More on this in Chapter 8, "Death of a Spacecraft."

As the amount of detail and documentation increases with each development phase, so do the number of people involved. A Pre-Phase A study may be carried out by a design team of 20 people, but when the project arrives in Phase C/D, hundreds of engineers, technicians, scientists, software programmers, documentalists, managers, secretaries, planners, contract officers and lawyers can be involved.

For instance, during Pre-Phase A the preliminary design of the electrical subsystem may be carried out by one single expert. In Phase B the details may be carried out by a Solar Array engineer, a Battery expert, a Power Control and Distribution engineer, and a Harness designer who takes care of all the cabling needed to connect everything. In Phase C/D the Solar Array, Battery and the rest of the equipment are each designed, built and tested by specific contractor teams. Several hundred people may be involved in the project by then.

Several times during the development and at the end of each official phase, the work performed until then is presented to the responsible space agency. The agency's technical, scientific and administrative experts will then go through all the available documentation to check that everything is in line with the specifications and that all the right analyses and tests have been done in the right way. Only when all the agency's experts are happy with the results of the review is the project ready to move on to the next development phase.

A general problem for the designers and reviewers is that "if it looks good, it flies good," a maxim that more or less works in aircraft design is in no way valid for spacecraft. Due to their un-aerodynamic, blocky shapes with sharp edges and projecting antennas, space probes are seldom pretty. Only a deep and careful analysis can reveal whether a space probe is well designed.

Such extensive reviews are a bit like exams for the spacecraft designers. However, they are not really intended as tests, but primarily as a means of finding problems and inconsistencies in the concept. The spacecraft design can then be improved before it is too late. Space probes are such complex machines that it is hard not to overlook anything, and space is a particularly unforgiving environment. A critical review by people who are not too involved in the project, and are therefore reasonably unbiased, is often crucial for its success.

A classic illustration of what can happen if reviews and checks do not root out serious problems is that of NASA's Mars Climate Orbiter, which arrived at Mars in September 1999.

Using its main rocket engine for braking, it began to slow down to attain an orbit around the red planet. While doing this, the probe disappeared behind Mars as planned. However, when the spacecraft should have re-appeared there was only static noise on the radio-link. Mars Climate Orbiter was neither seen nor heard from again.

Subsequent investigations concluded that the spacecraft had flown too low, plunged deep into the Martian atmosphere and burned up due to the heat of the aerodynamic shock caused by the collision with the denser air. A faulty look-up table embedded in the onboard computer software, which by mistake used imperial instead of metric units (pounds force instead of newtons), was found to be the cause of the disaster.

This simple, but very important mistake should have been noted during the development, but was overlooked because of the meager budget available for such technical management and control issues. The cutbacks in spacecraft tracking from the ground, again to save money, meant that the problem was not even identified during the flight.

Another good (or bad) example of insufficient development oversight is the maiden flight of Europe's new Ariane 5 launcher. The onboard inertial reference system, which tells the rocket where it is going, incorporated software from the launcher's predecessor, the Ariane 4. However, the flight of an Ariane 5 is quite different from that of the older type of rocket. Specifically, the horizontal velocity early in the flight is much higher.

Due to the obsolete software, the inertial reference system thought it was inside an Ariane 4 instead of the much more powerful Ariane 5. When the horizontal velocity exceeded the pre-set Ariane 4 limits some 30 seconds after liftoff, both the main and the identical backup systems shut down. The launcher veered off course, resulting in extreme aerodynamic forces that broke the Ariane 5 apart.

The extensive reviews and tests carried out during the Ariane 5 development program did not include adequate analysis and testing of the inertial reference system. This was a serious and costly mistake, because the launcher was also carrying four expensive satellites intended to investigate the Sun and its effect on the Earth's magnetic field (fortunately, updated copies of these "Cluster" satellites were later built and launched with success).

If the initial plan for the mission can be viewed as the project's conception, then Pre-Phase A is the project's birth, Phase A resembles infancy and Phase B is its early childhood.

During Phase C/D the project goes through puberty: the design evolves into its final form and emerges as a grown-up, complete spacecraft. Through various tests the probe can get a taste of what real life in space will be like. However, when something goes wrong, its "parents" – i.e. the developing space agency and industries – are still there to help out.

With the launch the probe enters its independent adult life, and can start to fulfill its promise and make its creators proud. As the parents can still be contacted over the phone (or radio in this case), the probe can tell them how it's doing, and can send back pictures of the exotic places it visits. It can even complain about aching mechanisms and ask for advice through the remote uploading of new software. Physically, however, the probe is then on its own.

3

ANATOMY OF A SPACE PROBE

ROBOTIC space probes usually do not look like people or animals. Rather than resembling the human-looking science fiction movie robots C3PO and Data, they have more in common with the squat, cylindrical R2D2. Like this famous robot from *Star Wars*, our spacecraft normally do not sport arms, legs, eyes and ears as we do, but they nevertheless function very well in their own special environment.

Like living beings on Earth, each type of robotic space explorer is optimized for specific purposes. A swallow flies low, speeding and zigzagging between trees and village buildings, and snatches small insects out of the air. To enable it to do that, the bird's body is small and lean and it has short, narrow wings that allow rapid maneuvers. An eagle, on the other hand, is much larger and has long, broad wings with which it can soar high, and almost effortlessly, over vast open areas. Its eyes are powerful optical instruments that can spot a small deer or rabbit below. If it finds one, it dives down and employs its mass and long talons to grip its victim.

Swallows and eagles are both birds. They have very similar body

FIGURE 3.1 Even though they do not look like the stereotypical space robots seen in classical science fiction movies, space probes are robots nonetheless. [MGM]

layouts, lightweight skeletons and feathers. Nevertheless, a comparison shows important distinctions that betray their very different ways of living. It is similar for spacecraft; although they broadly have the same types of basic equipment on board, probes designed to slowly radar-map cloudy Venus look different from those that make snapshot pictures of comets during fast flybys.

For instance, probes exploring planets close to the Sun have relatively small solar arrays. With plenty of light, only limited numbers of solar cells are needed to provide sufficient electrical power. In contrast, spacecraft sent toward the outer regions of the Solar System require very large solar arrays, or even need to use completely different systems for electrical power generation when incident light levels are too low for solar cells.

If probes flying through open space are like different species of birds, then they differ from planetary landers as birds differ from fish. Many internal parts are the same, but nevertheless they look vastly different. Landers are compact and sturdy with legs and wheels; orbiting probes often have ungainly long, fragile antennas and solar arrays sticking out that would immediately break off if deployed on the surface of a planet.

Nevertheless, however much robotic spacecraft may differ, you will also find similarities if you look closely and deeply enough. Like animals and people, space probes need to know what is going on around them and be able to react to that. Spacecraft need to keep their own temperatures within comfortable limits to provide sufficient power to all their active parts and to withstand the various forces and pressures they encounter.

A space probe should also be able to move itself, with thrusters if it is in space or with legs or wheels if it is on a planet with sufficient gravity. Moreover, a spacecraft needs to communicate with its controllers back on Earth, and it needs a computer brain to be aware of its environment and direct all its actions.

Bits and Pieces

The various elements on board a space probe are organized into subsystems, which consist of equipment that can be further broken down into components. For instance, the electrical power subsystem consists of equipment such as solar arrays and batteries, which themselves are built up of such components as solar cells, solar array hinges and electrical wires.

Like in an organism, all the subsystems and equipments on board a spacecraft influence each other and thus cannot be developed and built completely independently.

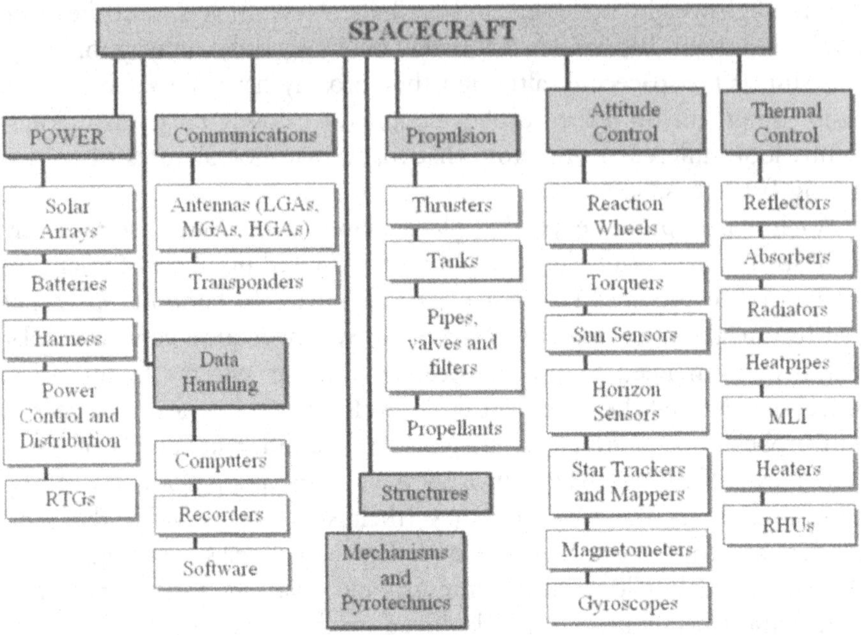

FIGURE 3.2 A typical breakdown of a spacecraft into subsystems and equipment.

For instance, the design of the power subsystem has a considerable influence on the amount of heat that is generated within the spacecraft (the more power goes around, the more heat the electrical equipment will give off), and thus has a strong effect on the thermal control subsystem. Likewise, the size and, thus, mass of the solar arrays has to be taken into account in the design of the spacecraft's supporting structure. If the solar arrays need to be larger, the structure will have to be adapted to handle the increase in size and mass.

Making things even more complicated is the fact that, to keep costs down, modern space probes often reuse as much existing equipment as possible. The use of tailor-made equipment optimized for a specific space mission is expensive and is thus often minimized.

This means that spacecraft engineers need to be able to put together new probes for new missions from a good deal of existing equipment, much like building an entirely new toy car out of the standard Lego blocks from the box of another model. The Lego blocks you would prefer to use may not exist, in which case you'll have to take blocks that look something similar to your ideal building blocks and adapt your design to suit them.

This approach was very successfully applied to ESA's Rosetta, Mars

Express and Venus Express missions. Specialized equipment developed for the large and expensive Rosetta comet chaser was reused for the smaller, low budget Mars Express and Venus Express spacecraft. The two Express probes may not have been ideally suited for their tasks, but they could be developed in a very short time and for a relatively low budget; Mars Express was developed for about 60 million euro (excluding instruments and Beagle 2 lander), but could easily have cost twice that amount if all equipment had to have been newly developed. When space science budgets are as low as they currently are, minimizing costs in this way can mean the difference between a less than optimized but flying mission and no mission at all.

Most people know more or less what a typical satellite or interplanetary spacecraft looks like, but have basically no idea of how they actually work. It's like the medical profession in the Middle Ages, when doctors were very familiar with the outside appearance of people, but had only a rudimentary idea of what was inside.

In the seventeenth century, scientists therefore started to dissect the bodies of dead people to find out what the internal organs looked like and how Mother Nature made all these parts work together. Often these scientific cutting sessions were held in small theaters, so that a larger audience of students and medics could learn the secrets of the human body.

We still cannot design anything as complicated as a person's body, but all the dissecting and probing has given us a pretty good idea of how it all works. Now, let's do something on that same principle. Rather than trying to design a spacecraft ourselves, we will put a typical planetary orbiter probe on an imaginary dissecting table and inspect its different parts.

POWER

Looking at the outside of a great many space probes, some of the first things you notice are the solar arrays with their shiny blue solar cells. On most spacecraft they come in the form of two large wings; on others the solar cells are simply covering the spacecraft's body all around. The thousands of interconnected solar cells work together to convert sunlight into enough electricity to power all the electrical equipment and instruments on the space probe. While people and animals need food and air, spacecraft need electric power.

The solar cells on space probes are not much different from those you may find on the roof of some houses, or those used to power your pocket calculator. The main difference is that those used in space need to be as efficient as possible to limit the size and mass of the solar arrays, and need to be protected with special cover glass to avoid damage by radiation.

To produce sufficient power – in the order of a few hundred watts for small space probes to several kilowatts for the big ones – solar arrays usually need to be fairly large. Sometimes they fit on the main spacecraft body, but often the number of solar cells is too high for that. Moreover, the arrays may constantly need to be pointed to the Sun to assure an adequate power supply.

This is why most spacecraft are equipped with deployable solar array wings. During launch they are folded close to the body of the spacecraft to make them fit inside the payload fairing of the launcher, but in space they are deployed like an accordion.

Once unfolded, Solar Array Drive Mechanisms at the roots of the wings are used to rotate them and keep them aimed at the Sun while the probe moves through space.

Sometimes there is a short period when the solar arrays cannot be pointed at the Sun – for instance, when the spacecraft needs to make a special maneuver or when it is in the shadow of a planet. A rechargeable battery is then needed to keep the electricity flowing until the Sun shines on the arrays once again. In any case a battery is needed to bridge the time between launch and the moment the solar arrays are deployed.

Just like solar cells, such batteries are basically not very different from equipment commonly used on Earth. What makes them so special and expensive is the fact that they need to be highly reliable over a long time, because they normally cannot be repaired or replaced once the spacecraft has been launched.

In the same way that veins and arteries transport blood to the various organs in your body, a so-called Electric Harness of power cables carries electricity from the batteries and solar arrays to the various power-demanding spacecraft equipment. A typical space probe contains many kilometers or miles of cables.

To ensure that the proper amounts of power get to the right places under the right voltages, a Power Control and Distribution Unit is included that works not unlike the way the heart manages the oxygen-transporting bloodstream flowing around your body.

When you demand more of your body than your lungs and heart can handle, you get tired and body functions start to shut down; your legs refuse to run anymore, and your arms can no longer lift those weights you

FIGURE 3.3 The two solar arrays on ESA's Mars Express were folded during launch and only deployed in space. [ESA/D. Ducros]

were pulling in the gym. Likewise, a spacecraft's onboard computer will need to shut down equipment when the need for power is more than the solar arrays and the battery can provide.

Power hungry scientific instruments are usually the first to be cut off, with the most vital equipment such as the onboard computer itself being partly put off only in extreme cases. Shutting down functions of the onboard computer is similar to your body shutting down parts of your brain in case of extreme fatigue: you faint to put your body and brain on low-power standby to prevent serious or permanent damage. Likewise, a spacecraft's computer may go to standby when power levels drop below a certain minimum.

Sometimes spacecraft are deliberately put on low-power standby. This is called "hibernation," a kind of winter sleep that lets the spacecraft survive on a minimum of electricity, like a bear sleeping through the winter without eating or drinking.

Hibernation is often used on an interplanetary spacecraft that needs years of traveling through space to reach its destination. While hibernating, the spacecraft does not need to be pointed at the Sun continuously and the inactivity minimizes wear and tear on the onboard equipment during the long travel. It also minimizes the amount of manpower needed in the mission control and spacecraft tracking facilities back on Earth. This helps to limit the cost of running the spacecraft for such long missions.

The more sunlight, the more electrical power a solar array of a certain size will be able to produce. That is why spacecraft moving closer to the Sun, like those sent to Venus or Mercury, normally require only rather small solar arrays.

However, at a certain distance from the Sun, even before the orbit of Mercury, the temperature of the arrays can get too high because of the Sun's heat. Instead of being perpendicular, the solar arrays then have to be pointed at an angle to the Sun. In this way less heat radiation hits the array surface, but there is also less useful light for the solar arrays. The arrays have then reached their highest possible output; moving closer to the Sun means that the angle with the Sun has to be lowered even more, so no further benefit from the increasing amount of light will be possible.

For probes flying outward from Earth and the Sun the situation is reversed. Spacecraft flying beyond Mars venture into the outer regions of our Solar System, where there is little light to enable the solar arrays to produce electricity. Instead of very large and heavy solar arrays, engineers can opt to equip such interplanetary probes with Radio-isotope Thermo-electric Generators, or RTGs.

RTGs are very compact and very reliable power systems. They work on the thermo-electric principle that a voltage can be generated between two different conducting materials if they are each kept at a different temperature. In an RTG the temperature difference is created by the heat emitted from the natural radioactive decay of plutonium (mostly non-weapons grade Plutonium-238), which is in the form of ceramic plutonium dioxide.

Because RTGs produce a lot of heat and radiation that can disturb other equipment, they are usually kept as far as possible from the main spacecraft body by putting them on long poles on the outside.

With RTGs a space probe does not need sunlight or any other external source of energy. Apart from enabling missions to the dark outer reaches of the Solar System, they can also be used at places where the radiation would damage solar arrays. That makes RTGs ideal for use on a future

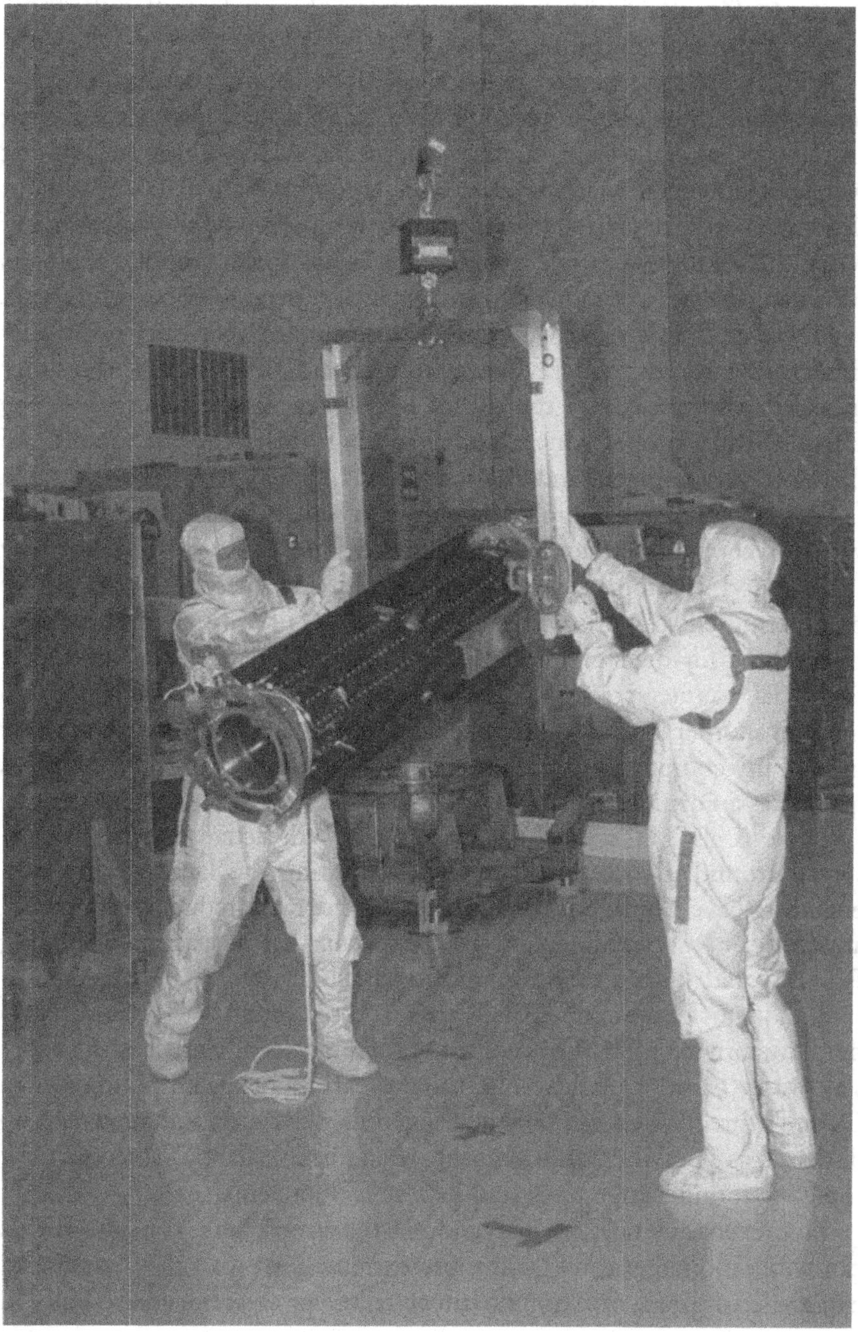

FIGURE 3.4 One of the RTGs used on Cassini just before integration with the spacecraft. [NASA]

orbiter to, for instance, Io, the volcanic moon that circles within the fierce radiation zones around Jupiter.

RTGs would also be very practical on probes landing on planets where long nights, thick clouds and dust in the atmosphere prevent sufficient amounts of sunlight from reaching a spacecraft. The Viking landers that were put down on Mars in 1976 were equipped with RTGs. Without depending on solar cells that would slowly be covered by the dusty sand of Mars, the Vikings were able to operate for years. Future robotic explorers on Titan, the veiled moon of Saturn, may also need to depend on them.

However, RTGs also have a serious disadvantage: they carry potentially dangerous amounts of highly radioactive material. The safety and security measures associated with the use of RTGs on a spacecraft such as the Cassini Saturn probe, which is now touring Saturn and its moons, are therefore very stringent and elaborate.

First, technicians handling RTGs have to be very careful, and also work very quickly because they are only allowed to be close to RTGs for very short periods. According to European health and safety rules, people are only permitted to work with unshielded RTGs for a maximum of two minutes ... for their entire life! That does not leave a lot of time for a technician to integrate an RTG into a spacecraft of course; after just two minutes another person will have to take over. RTG integration teams test on dummy equipment until they can install the dangerous units nearly as quickly as a Formula-1 pit stop team can change the tires on a racing car.

There is also, of course, the chance that radioactive material will be spread around if something goes wrong during the launch. However, engineers have been very successful in considerably reducing the risk of any radioactive contamination in the event that a spacecraft carrying RTGs would suffer a catastrophic launch failure.

Unlike nuclear reactors, RTGs are not based on nuclear fission and do not have moving parts. They consist of solid blocks of material that simply stay warm because of their internal nuclear energy. The plutonium in Cassini's three RTGs can only be dangerous if it would somehow end up dispersed as a powder in the atmosphere and be inhaled by someone. Or if someone found a large piece and kept it as a souvenir.

To prevent any possible dispersion in the atmosphere, Cassini's RTGs can withstand larger forces and more extreme heat than they could ever experience during a catastrophic launch failure or an accidental re-entry in the atmosphere.

The first safety measure is that the radioactive fuel on board Cassini is in the heat-resistant, ceramic form of plutonium dioxide, which greatly reduces the chance of the material evaporating in a fire or during a re-entry.

Next, the fuel is divided into 18 independent units, each of which has its own heatshield and impact shell to avoid the fuel from evaporating during re-entry or breaking into pieces when falling on the ground. Dividing the fuel into small units almost eliminates the chance that all the radioactive material could somehow end up in the environment (the downside is that all this separate packaging makes the RTG much heavier).

Finally the individual fuel units are packed together in a shell with multiple layers of strong, heat- and corrosion-resistant protection materials. The RTG is designed to survive any disaster intact, so that it can be recovered without causing any problem to anyone.

The USA has flown 23 missions with RTGs over the last 40 years, and they have seldom caused any problems. In 1964 an RTG with 1 kilogram (2 pounds) of plutonium burned up while descending through the atmosphere after a failed launch. All the radioactive material was vaporized and spread throughout the atmosphere. After this accident it was decided to build future RTGs in such a way that they would remain intact when falling back to Earth. In 1968 two RTGs came down after another launch failure, but they were recovered intact and actually reused on another satellite.

The Russians have a less satisfactory safety record with their RTGs. In 1978 a defunct radar spy satellite broke up over Canada, spreading 50 kilograms (110 pounds) of dangerous Uranium-235 chunks over a large part of the country. The US and Canadian authorities spent a couple of weeks searching 47,000 square kilometers (18,000 square miles) under harsh winter conditions to find all the fragments. In addition to this disaster, Soviet launch failures and re-entering satellites have been responsible for large amounts of radioactive material evaporating in the atmosphere.

However, the chance that even the tiniest amount of plutonium from modern RTGs, such as those in Cassini, could ever end up in the atmosphere is remote. Moreover, in the extremely unlikely event that any plutonium does escape, the particles in the atmosphere would also have to be inhaled by someone and stay in that person's body for a long time to cause any danger. Even then, it is most likely that the amount of any such particles inhaled would still be far too small to cause any measurable damage.

Because of the special measures taken, the risk that RTGs pose to your health during a launch is less than the risk that you could be hit by lightning or even a meteorite from space – and the chances of that happening are very low. In 1954, a woman in Alabama, USA, was actually struck by a meteorite. The 9-pound rock crashed through her roof, bounced off a radio, and hit her while she was sleeping on her couch. Luckily, because the stone ricocheted off several objects before striking her, she only had some bruises. More recently, In August 1992, a big meteorite disintegrated over Mbale, Uganda. A small fragment was slowed by the foliage of a banana tree before falling on a boy's head. Fortunately he was not injured.

In spite of the extremely low public health risks involved, Cassini's RTGs made it the most vigorously opposed and defended spacecraft ever. Groups of anti- and, in reaction, pro-Cassini activists grew rapidly as the 1997 launch got near. Angry debates were held on websites, in newspapers and on television. One anti-Cassini protest at Cape Canaveral even resulted in the arrest of 27 activists who tried to trespass on the launch base.

In spite of the protests, the launch campaign continued unhindered, and Cassini was sent on its way as planned. However, the fight was not over yet, because the probe's trajectory included an Earth flyby in 1999, to pick up enough speed to make it to Saturn. This time anti-Cassini activists feared that the probe would inadvertently end up and burn in the atmosphere, releasing plutonium from its RTGs in the process. They put pressure on NASA and the government to abort the flyby and put the space probe on a trajectory away from Earth, but again their protests went without effect.

The Earth flyby did happen as planned, increasing Cassini's speed by 5.5 kilometers (3.4 miles) per second. This was an important part of the total 21.4 kilometers (13.6 miles) per second it needed to accumulate during its four planetary flybys (twice past Venus, once past Earth and once past Jupiter). Without the critical Earth flyby Cassini would never have made it to Saturn, where it is now making amazing discoveries and returning beautiful pictures of Saturn's moons and rings.

Both RTG and solar cell technology is continuously developed further. There are plans to integrate RTGs with equipment that does not use the heat produced to make electricity directly, but instead employs thermodynamic or chemical processes to run electricity generators. This can result in much higher efficiencies, so that less radioactive material is required in the spacecraft.

At the same time, solar cells are developed that can operate with less light than normal. This will make it possible to use solar arrays as far away from the Sun as Jupiter. We may see Jupiter probes in the near future that will not need any RTGs.

COMMUNICATION

Space probes are not alone in the Universe. They are part of a "society" of engineers and scientists without whom the machine's life is meaningless. Interplanetary space probes need to be able to talk to us, to tell us how they are and send us the streams of data from their scientific instruments. Likewise, people in the mission's control room need to talk to the spacecraft to make it change its orbit, to switch on a certain instrument and point it at an interesting object or to help the probe to recover from unforeseen equipment failures.

Contact with our robots in space is maintained via radio. Large ground-based parabolic dish antennas are used to sufficiently focus the energy of our transmissions so that they can be received by spacecraft hundreds of millions of kilometers distant.

FIGURE 3.5 The Canberra ground station antenna dish of NASA's Deep Space Network for communication with interplanetary spacecraft. [NASA]

Space probes cannot carry such large antennas as that would make them impossible to launch. The signals they send us from deep space through their smaller antennas are extremely weak by the time they arrive on Earth. What we receive from the Cassini orbiter at Saturn for instance, is about 20 billion times weaker than the energy output of the tiny battery in a digital wristwatch. However, the same large ground-based antennas we use to send powerful signals can also work the other way around. Their huge dishes are used to collect as much of the diluted spacecraft radio messages as possible, and focus that data into a strong enough signal for our receivers to detect it.

The type of communications equipment a space probe carries depends on the amount of data it has to send, the distance from Earth it will travel, the amount of available electric power on board, and equipment mass limitations.

Antennas are the ears of a spacecraft. They are divided into three main groups: high-gain antennas (HGAs), medium-gain antennas (MGAs) and low-gain antennas (LGAs), comparable to the huge sound-collecting surface of the ears of a coyote, the medium-sized ears of a horse and the very small ears of a duck. However, unlike biological ears, spacecraft antennas are also used to send out signals. Spacecraft antennas are like ears and mouths, and these functions are often combined in the same antenna.

The "gain" achieved by an antenna is a measure of the relative amount of incoming radio power it can collect and the strength of the signals it can transmit to Earth. The higher the gain, the stronger the signals it can put out and the weaker the signals it is able to detect.

When you look at the outside of an interplanetary space probe, the HGA is one of the things that will catch your eye. HGAs are dishes that look a lot like the giant parabolic antennas we use on Earth, but they are much smaller. Nevertheless, they are often the largest thing on the spacecraft.

Like optical telescopes, HGAs can focus the radio waves they collect or send out, but their field of view is very narrow. For HGAs to work properly, they have to be accurately pointed to Earth. Fixed HGAs make it necessary to maneuver the whole spacecraft to aim the antenna. Other space probes have an HGA that can be independently steered with a mechanism, but this comes at the price of higher equipment mass and added complexity, of course.

Galileo, the orbiter that explored Jupiter, had an HGA that consisted of a 5-meter (16-foot) diameter mesh stretched over 18 curved ribs. The benefit of this design was that the whole contraption could be folded up like an umbrella during launch on board the Space Shuttle, which enabled

FIGURE 3.6 The large dish-shaped High Gain Antenna on ESA's Ulysses was used to communicate with the spacecraft in its orbit over the Sun's poles. [ESA/NASA]

the use of a larger diameter antenna than would normally have fitted in the Shuttle Orbiter cargo bay.

However, when almost 18 months after launch NASA controllers tried to deploy the HGA, something went wrong. The procedure, tested on Earth years earlier, called for a set of electrical motors to drive the antenna ribs out from a central mast and thereby open the umbrella. Once fully deployed, the ribs would snap into place and Galileo would send a notice of success to the ground. It didn't.

Within minutes after sending the deployment command, Galileo's flight team saw that, although electricity was flowing through the deployment motors, they were not turning. The antenna, crucial for transmitting the expected huge amounts of scientific data from Jupiter, was stuck.

More than 100 people quickly got busy with testing, simulating, analyzing, consulting and reviewing, in order to find a way to get the antenna unfurled completely. They discovered that three or four of the antenna ribs suffered from friction between misaligned standoff pins and their sockets, and therefore got stuck to the central tower of the antenna.

The first remedial action consisted of turning the antenna toward the Sun to warm and expand the central part of the antenna, which would hopefully free the stuck pins. When this didn't work, they tried

FIGURE 3.7 The astronauts' patch for shuttle mission STS-34 that put the Galileo spacecraft on its way. [NASA]

repeated thermal cycling: warm–cool–warm–cool. However, as the spacecraft was quickly receding from the Sun, warming had a limited effect and the temperature could only be raised by about 30 degrees Celsius (50 degrees Fahrenheit).

That was not enough; Galileo had to fly on with an antenna that was only partly opened and therefore useless. If someone could have gone out to the spacecraft, the problem would have been fixed in no time. However, once an interplanetary probe is on its way, it is too late for even the simplest repair.

Later the cause of the problem was traced back to Galileo's transportation to the launch site on a flatbed truck. Bouncing its way along thousands of kilometers of uneven, potholed freeways, pressure exerted on one side of the packed antenna probably caused the lubricant on a few rib pins to wear away. When Galileo reached the launch site in Florida, no one had checked the lubricant.

With the HGA permanently stuck, Galileo had to carry out its mission using a low-gain antenna. This meant that the sending of images and other scientific information from its instruments was constrained to low data rates. Fortunately, by remotely programming Galileo's computer with new ways of compressing the data, and by being patient, most data nevertheless eventually found its way to the scientists on Earth.

HGAs are sometimes creatively used for other purposes. The dish HGAs on NASA's orbiters Magellan and Cassini doubled as sunshades when the spacecraft were close to Venus and needed to be pointed at the Sun. Cassini's HGA also served as a protection shield against the thousands of micrometeoroid impacts it endured when, upon arrival at Saturn, it had to cross the planet's ring plane. It flew through a gap between the main rings, which consist of dust and ice particles.

If we compare HGAs to optical telescopes, then low-gain antennas can be compared to the naked eye. LGAs cannot collect and focus as much data as HGAs, but because they have a much larger field of view they don't need to be pointed very precisely.

An LGA only allows low data rates, but it can be used before an unfoldable HGA is deployed, or when the spacecraft is for some reason unable to point its HGA in the direction of Earth.

LGAs are small and light, so spacecraft usually carry at least two, one for redundancy in case the other does not work properly, and to ensure that at least one LGA is always able to have an unobstructed link with the Earth.

Medium-gain antennas are a middle solution. MGAs provide more gain than LGAs and allow wider angles of pointing than HGAs. The Venus-exploring Magellan spacecraft for instance carried a large cone-shaped MGA, which was used during some maneuvers when the HGA could not be pointed at Earth.

The antennas on the outside are linked to the radio receivers and transmitters you can find inside the spacecraft. Modern receiving and transmitting equipment is usually combined into single units called "transponders." Transponders do not just receive and transmit on command, they also actively listen for and respond to signals coming from Earth.

If a spacecraft can no longer transmit or receive radio signals, it will become utterly useless. Therefore satellites and interplanetary spacecraft usually carry at least two transponders – one for normal use, and one as a backup in case the other breaks down.

Radio links with interplanetary spacecraft can be maintained over incredible distances. Until recently we could still talk with Pioneer 10, although it is now an incredible 12.3 billion kilometers (7.6 billion miles) from Earth.

NASA's Pioneer 10 was launched on a powerful Atlas–Centaur rocket on March 2, 1972. As the fastest human-made object ever to leave the

Earth, it became the first probe to reach Jupiter as it flew past the largest planet in the Solar System on December 3, 1973.

The spacecraft was the first to obtain close-up images of Jupiter, charted the gas giant's intense radiation belts, located the planet's magnetic field, and proved that Jupiter is predominantly a gas planet.

Originally Pioneer 10 was designed only for the 21-month mission to Jupiter, but thanks to its rugged design and the continuous power supply by its RTG, Pioneer 10 lived on. It continued to explore the outer regions of the Solar System, studying energetic particles expelled by the Sun (the solar wind) and cosmic rays coming from outside the Solar System.

In 1983 it became the first human-made object to pass the orbit of Pluto, the most distant of the original nine planets (we now know there are more Pluto-like objects further out), and headed into the great dark unknown. The sturdy Pioneer continued to send valuable scientific data via its huge HGA until March 31, 1997, when the RTG's output became insufficient to power its instruments.

Nevertheless, the basic, weak "I'm here" signal from Pioneer's transmitter could still be received on Earth. At 82 times the distance of the Earth to the Sun, even traveling at the speed of light, the space probe's radio signals took 11 hours 20 minutes to reach the Earth.

The attempts to receive Pioneer's signal had by then become part a new study of large-distance communication technology. The knowledge gained was used in support of NASA's future Interstellar Probe mission that is now on the drawing boards (see Chapter 9: "A Bright Future").

Finally, on January 22, 2003, Pioneer 10's useful life came to an end when its last, extremely weak signal was received. More than 30 years after it had left our planet.

"It was a workhorse that far exceeded its warranty, and I guess you could say we got our money's worth," commented Pioneer 10 Project Manager, Dr Larry Lasher, when Pioneer 10's long mission was finally over.

STRUCTURES

All the things you see sticking out of a spacecraft, and all the things inside that you cannot see, need to be attached to a main structure, a skeleton that holds everything together.

A spacecraft structure needs to be large enough to house and protect all the equipment and be strong enough to ensure that the spacecraft does not

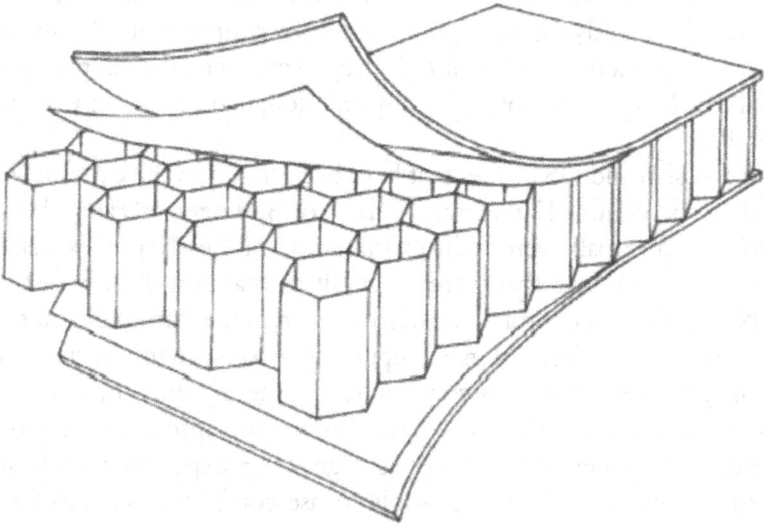

FIGURE 3.8 Aluminum honeycomb panels are light and stiff, making them ideal for spacecraft structures. [NASA]

fall apart during launch and operation in space. It must also allow sufficient access for technicians to fit in all the equipment. Moreover, it needs to do all of this while being as light as possible.

Modern spacecraft structures are usually made of panels consisting of an aluminum honeycomb section sandwiched by aluminum sheets. These panels are very strong and stiff, but also very light and therefore ideal for spacecraft construction.

The Russians liked to apply titanium as well, not only because it is strong and lightweight but is also able to handle much higher temperatures than aluminum. However, titanium is much more difficult to cast and machine; drill bits designed for piercing aluminum sheets simply break into pieces if you try to drive them through titanium, therefore special tools and manufacturing facilities are needed to build titanium structures.

It is also very difficult to extract from mined titanium-containing minerals, and therefore, in general, titanium is rather expensive and not easy to obtain. However, during the Cold War, the Soviet Union was one of the few nations with rich titanium mines, so making spacecraft structures out of this precious metal was for them less of a problem than it was for other spacefaring nations.

Because of their own very limited reserves of titanium, the USA even had to buy from the Soviets the large quantities of titanium needed for the super-secret SR-71 Blackbird plane. The Blackbird needed to be built from titanium because of the high aerothermodynamic temperatures it encountered when flying through the air at over three times the speed of sound. The Soviets never realized they were actually supplying the material for the construction of an airplane designed to spy on their own territory!

Most Russian and US probes that have been sent into the extremely hot atmosphere of Venus (470 degrees Celsius or 880 degrees Fahrenheit on the surface is normal) were built of titanium. The early probes that the Russians sent to Venus that were not built of titanium did not last very long. NASA's Galileo probe, which dived into the turbulent, hot and high-pressure atmosphere of giant Jupiter, also had a titanium structure.

Advanced materials such as composites – some of which may be a mix of plastics, metals and glass fibers – are also used on spacecraft. Previously only employed when really necessary, growing experience with such novel structures now makes it possible to use composites such as CFRP (Carbon Fiber Reinforced Plastics) more frequently.

Often all kinds of beams, tubes, trusses and baffles need to be added to the main structure to ensure that all the various equipment can be mounted in and on the space probe.

The driving requirements for the design of spacecraft structures are usually not set by the harsh space environment in which the probe will operate year after year, but by the launch phase that may take only a mere 10 minutes. During launch, the spacecraft is accelerated, shocked and shaken by the rocket it rides on. The launcher's rocket engines and the tremendous noise of the launch and aerodynamic buffeting during flight cause powerful and potentially damaging vibrations, like music that is too loud in a glassware shop.

Moreover, the jettisoning of empty rocket stages, the ignition of the powerful rocket engines of subsequent stages and the final separation from the launcher's last stage cause sudden shocks that can really kick a spacecraft where it hurts.

Probes that are meant to land on planets often also experience violent shocks on arrival. NASA's Pathfinder and MER landers depended on airbags for their final fall to the Martian surface. The airbags and the landers were designed to expect a maximum deceleration shock, equivalent to 40 times the lander's weight on Earth!

Spacecraft structures therefore need to be carefully designed and tested. Fortunately, today's computer modeling technology has replaced much of the exhaustive trial-and-error testing and design sequences of the past. We can now efficiently optimize the structure for maximum strength and stiffness while keeping the mass as low as possible.

THERMAL CONTROL

Most equipment on board a spacecraft operates best at temperatures between –5 and +40 degrees Celsius (20 and 120 degrees Fahrenheit). However, in Earth orbit the part of the spacecraft in full sunlight can reach temperatures around 120 degrees Celsius (250 degrees Fahrenheit), while the shadow side may get as cold as –160 degrees Celsius (–260 degrees Fahrenheit). The temperatures can even soar to much higher levels if the probe gets closer to the Sun, for instance if it flies to Mercury.

In space, a spacecraft receives heat directly from the Sun, but also by energy reflected and radiated by the planet it may be orbiting (especially high if that planet is Venus). Moreover, a space probe generates heat itself through all its electrical equipment (and this can be a lot; just feel how much heat your television or CD player dissipates).

To prevent the temperature from getting too high, a spacecraft must be able to dump its excess thermal energy. Likewise, when it gets too cold a space probe must be able to heat itself.

The simplest way to control the temperature is by the use of passive thermal control systems that do not require any additional power. Spacecraft are therefore commonly covered with various metallic panels, thermal blankets, black or white areas and mirrors.

These are all used to manage the heat that is emitted by the electrical equipment inside the spacecraft. If it gets too hot, the passive systems need to be able to let heat out. If it gets too cold, the passive systems need to keep more thermal energy inside. Like the skin of an animal, the passive coverings need to maintain the right thermal balance – to trap just enough heat to stay comfortably warm independent of the outside environment. It's something similar to a good sleeping bag that keeps you warm on a cold night and cool when it is hot.

Solar reflectors reflect and emit infrared (heat) radiation; they prevent solar radiation from heating up the spacecraft surface, while letting excess internal heat escape. A good solar reflector is, for instance, a mirror or white paint (likewise, it is better to wear white clothes on a hot, sunny day instead of dark ones, and white cars stay cooler in the Sun than dark-painted cars).

So-called flat reflectors like aluminum paint also block solar radiation from heating up the spacecraft, but contrary to solar reflectors they prevent internal heat from leaving the space probe body. They stop heat from being transported through the spacecraft's outer panels.

Solar absorbers such as polished aluminum sheets soak up solar radiation and emit little energy, and thus help to keep a spacecraft warm.

Flat absorbers like black paint absorb solar energy as well, but also emit internal energy easily; they make a surface heat up quickly but also cool down quickly (for example, wearing black clothes in the desert makes you overheat during the day and freeze at night).

Radiators are another type of passive system. Unlike the radiators of the central heating system in your house, spacecraft radiators are actually used to cool down a space probe. They absorb excess heat from inside the spacecraft and radiate it into space.

Radiators are commonly used in combination with heatpipes, which are basically tubes containing a fluid. On the hot side – for instance, near an active electronics unit – the fluid in the pipe evaporates. Capillary pressure then makes it travel toward the radiator on the outside of the spacecraft where it condenses, giving off heat to the radiator that subsequently dumps it into space as infrared radiation. (Your radiator at home does the same, but in addition heats up air through conduction, which of course does not happen in the vacuum of space.) It's the spacecraft equivalent of sweating, with the difference that the fluid is not

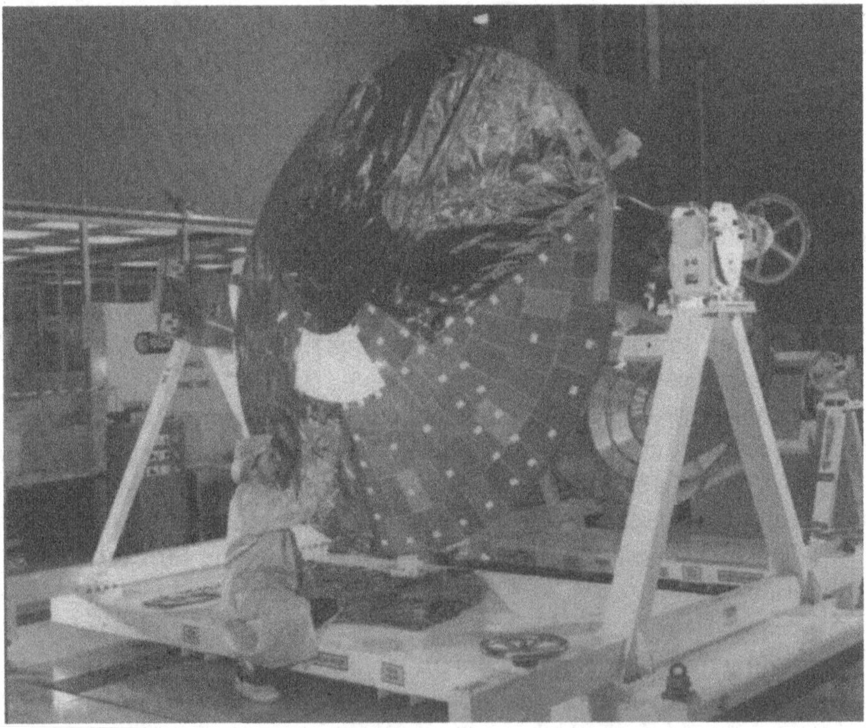

FIGURE 3.9 The heatshield of ESA's Huygens probe being covered with Multi Layer Insulation blankets. [ESA]

lost but can be used again. The condensed fluid travels back to the hot zone to absorb heat and evaporate, over and over again, in a continuous flow.

Heat pipes can thus be used to transport heat from inside the spacecraft to the outside. Even though the fluid is traveling back and forth, heatpipes work on thermal energy alone and do not need electricity.

Yet another way of passive thermal control is the use of insulation. MLI, for *Multi Layer Insulation*, is the crumply, golden sheets you commonly see on the outside of satellites (see Figure 3.9).

It consists of blankets built up from multiple separate sheets that alternate low-emitting surfaces (hardly emitting heat) with low-conductive barrier layers (hardly absorbing and transporting heat). Heat has a very difficult time getting through all these non-cooperating sheets, in both directions. Much like coffee or ice-water in a thermos flask, the warm items in the spacecraft stay warm, the cold items stay cold.

Covering a spacecraft with MLI is a real handicraft, as pieces have to be cut and folded in the right shape to fit the various surfaces, corners and protrusions. A spacecraft assembly room during MLI attachment somewhat resembles a children's origami class.

Sometimes a thermal control engineer can mix absorbers, reflectors, radiators and insulation in such a way that a good temperature balance can be reached within the spacecraft for its entire mission. An ideal space probe keeps itself warm by the heat its electrical equipment generates and some of the solar radiation it receives, while any excess heat is dumped into cold space by black surfaces and radiators.

Usually, however, passive systems are not enough. Active thermal control systems such as electrical heaters and sometimes even coolant fluids pumped through the spacecraft are then needed to enforce the right thermal balance.

Electric heaters work much like an electric blanket. Each heater consists of a simple wire that, because of its electrical resistance, generates heat when a current is passed through it. Thermostats similar to those used in your home switch the heaters on when the temperature drops below a certain minimum.

However, just as in people and animals, heat generation or cooling costs energy – electric energy in the case of spacecraft, resulting in the need for larger and heavier solar arrays and batteries.

Space probes landing on Mars or other planets often need to survive long, cold nights during which there is no light for the solar arrays to use to make electricity. To keep sufficiently warm without draining the rechargeable batteries, these probes can be equipped with Radio-isotope Heater Units (RHUs).

The RHUs on both NASA's Mars Exploration Rovers, Spirit and Opportunity, consist of 2.7 grams of radioactive plutonium-238-dioxide, contained in a platinum–rhodium alloy cladding. All together an RHU weighs about 40 grams. Each of the eight units on board each rover emits about 1 watt of heat, simply by the radioactive decay of the material.

They need no power or control system and can easily be placed where the spacecraft's other thermal control equipment has a hard time maintaining the temperature.

Nevertheless they are not very popular, as apart from heat these little things also emit rather unhealthy levels of radiation. They can therefore only be inserted in the nearly complete spacecraft just before launch, when no-one needs to be close to it any longer. All handling of the RHUs has to be done according to very stringent safety and security rules.

Moreover, before launch it must be guaranteed that if the rocket fails the RHUs will either safely disappear into the depths of the ocean (where they are supposed to be harmless) or can be retrieved intact and removed from our environment. Each RHU of about the size of a camera film roll therefore consists mostly of shielding, to prevent the release of radioactive material in the environment in case of a launch failure. The plutonium fuel pellet itself is only about the size of a small pencil eraser.

RHUs can survive rocket explosions, fires, the intense heat of an accidental atmospheric re-entry and the impact of hitting the ground from a high altitude. They have been tested in far worse conditions than they could ever encounter during a real launch, and have proven to be extremely rugged and reliable.

However, another disadvantage is that RHUs cannot be switched off; they always produce heat, whether it is needed or not. That means that the spacecraft may actually require a cooling system to remove excess RHU heat during the "warm phases" of its mission.

Thermal engineers and power subsystem designers always have to work closely together during the development of a new space probe. A higher electricity requirement results in a higher (solar) power generation, which means an increased internal thermal dissipation. This usually means trouble for the thermal engineer, who has to design a thermal subsystem that will remove the excess heat.

However, when the spacecraft is in the shadow of a planet and the spacecraft cools off rapidly, the electrical power expert has to make sure that the batteries contain enough power to keep the inside of the probe comfortably warm.

ATTITUDE CONTROL

A spacecraft needs to control its own attitude while flying through space. This is necessary to point the scientific instruments to the target, the solar arrays to the Sun, the antennas to the Earth, and to keep the heat-dissipating radiators in the shadow. Moreover, before firing a rocket thruster to do orbit maneuvers, the nozzle has to be pointed in the right direction.

There are two main ways to control the attitude of a space probe. The simplest is to make the spacecraft spin. Like a spinning top, the gyroscopic effect will keep the rotational axis of the probe in the same direction, as long as the craft is in balance.

The spin can be started by a mechanism on the launcher or mother spacecraft just before the probe or capsule is separated from it, or by small rocket thrusters on board the probe itself. Once started, a mechanical force or rocket thrust is no longer necessary. As there is no friction with any atmosphere in space, a rightly balanced space probe that is set to spin will rotate at the same speed almost indefinitely (in reality its spin will be slowly disrupted by movements of equipment inside the spacecraft and the minute drag of the upper atmosphere while in Earth orbit). Small pulses of rocket thrusts are only needed occasionally to point the spacecraft in a different direction.

Spin stabilization is very efficient, as you hardly need any power or rocket thrust to maintain the spacecraft's attitude. Because of this simplicity and efficiency, most early satellites and interplanetary probes used spin stabilization.

The downside of spinning is that all the onboard equipment is also rotating, making it difficult to point cameras, antennas and solar arrays in a certain direction. Some therefore have scientific instruments installed on complicated de-spun platforms that are not rotating with the rest of the probe, so that cameras and other sensors can aim constantly at their targets.

Moreover, for a stable rotation a spinning spacecraft needs to be mass-symmetrical in the plane in which it is spinning, otherwise it would wobble. Spinning spacecraft are therefore usually cylindrical, with a horizontally placed antenna dish bolted on top.

To ensure that the solar cells are always in sunlight, they need to be mounted all around on a spinning spacecraft. This means that you need a relatively high number of solar cells, because while spinning only about one-third are sufficiently aimed toward the Sun at any time. If you need a lot of power, you may have to make your probe very large to fit all the necessary solar cells.

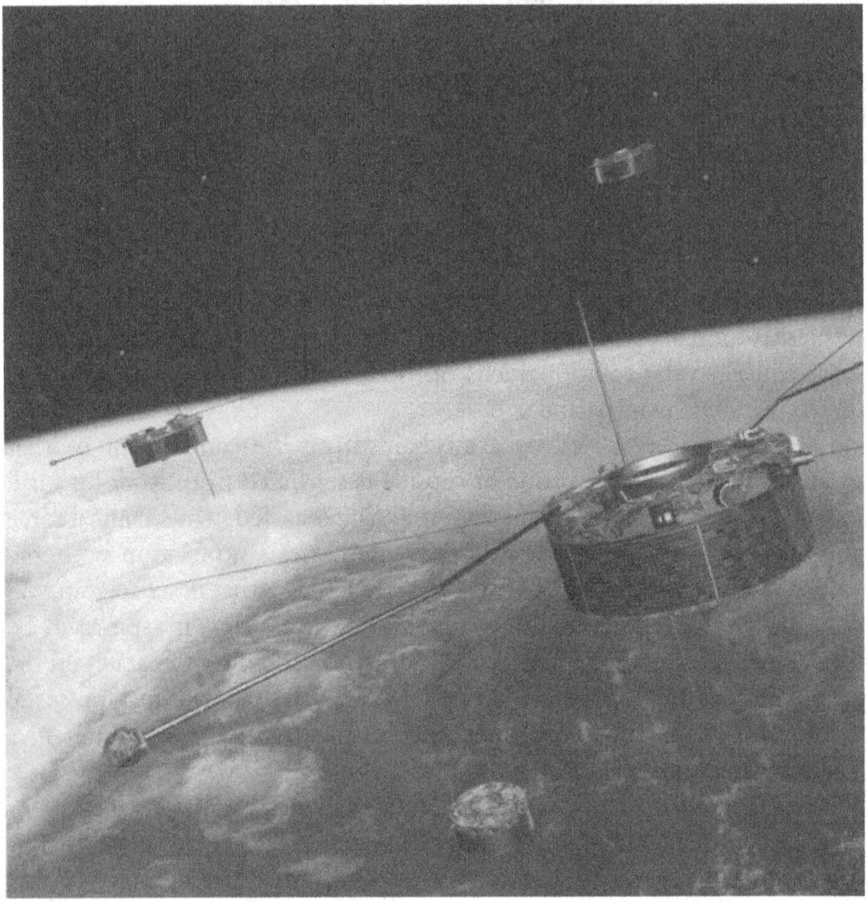

FIGURE 3.10 ESA's Cluster spacecraft are spin stabilized and work together to investigate the Sun's radiation and its effect on the Earth. [ESA]

All of this puts severe limitations on the design of spinning probes, although the method is still applied when the advantages outweigh the disadvantages.

A good example of the modern use of spinning space probes is ESA's Cluster II mission, which involves four identical, drum-shaped spacecraft. (The original Cluster mission was lost with the failed launch of the first Ariane 5 rocket described in the previous chapter, but new spacecraft were built and launched, and the project rose like a phoenix from its ashes.) The Cluster probes fly in formation in a very elongated Earth orbit through space, to study the flow of electric particles expelled by the Sun (the solar wind).

The four 2.9-meter (9.5-foot) diameter satellites roam the area where the solar wind crashes into the Earth's magnetic field, some 60,000 kilometers (37,000 miles) out on the sunny side of our planet. The Earth's magnetic field, the same that makes a compass turn north, acts as a shield to prevent the particles from reaching our atmosphere. It is much like the way a magnet can be used to repel other magnets.

We are lucky for this protection, as the electric particles can severely upset electrical equipment; sometimes a solar storm is strong enough to puncture the Earth's magnetic defense, producing beautiful auroras where the particles clash with the atmosphere, but also triggering power blackouts. Moreover, when the flow of charged particles hits the upper atmosphere, it heats it up and may thereby have an effect on our weather and climate. Furthermore, the atmosphere expands when it warms up, increasing the aerodynamic drag on low orbiting satellites and making them fall back to Earth much earlier than planned.

For scientific instruments that need to measure magnetic fields and particle streams, bolting them on spinning satellites like those of Cluster II results in a very beneficial sweeping motion. This ensures measurements of local averages rather than measurements at single, perhaps abnormal, points. The Cluster satellites have electric sensors whirling on 50-meter-long (160-foot-long) wires and magnetometers on deployable booms to cover a wide area.

An instrument on a spinning spacecraft also has a 360-degree field of view. That may mean that you need only one camera while, on a similar but non-spinning probe, you may need two or more.

The use of four satellites, as on Cluster, makes it possible to tell whether a certain change in the measurements is caused by the passing of time or by the movement of the spacecraft through space. With fewer spacecraft this would have been almost impossible to determine.

Capsules with instruments or landers on board that need to descend through planetary atmospheres are usually also spin-stabilized. With the right design, a spinning entry probe will always maintain its right attitude, even while crashing through the Martian atmosphere at tens of kilometers or miles per second. This is extremely important, as the thick heatshield that protects the probe against the superheated atmospheric gasses must be kept in front, and the parachute pack must be at the back. If the probe started to tumble, it could easily burn up, or its parachutes may wrap themselves around it.

However, most modern spacecraft are three-axis stabilized, which means that they do not spin but their attitude is actively controlled around all three possible rotational axes. This offers a lot more freedom in the design of the spacecraft and makes it possible to point the equipment in

FIGURE 3.11 A view at the round heatshield that protected ESA's Huygens probe during its fiery entry in the atmosphere of Saturn's moon Titan. The tiles are all the same; their different colors are a result of the production process. [EADS Space]

the most optimal directions. Three-axis-stabilized spacecraft are usually box shaped and, because they don't need to bother with spinning stability, can be very asymmetrical.

Nearly all three-axis-stabilized space probes have a reaction control subsystem consisting of a series of small thrusters distributed over the spacecraft. Firing a thruster on the left side and another in the opposite direction on the right side causes a rotation, as long as the thrust is not pointed exactly through the probe's center of mass. With six small thrusters a simple spacecraft can be nudged around all of its three axes.

Another method for three-axis stabilization is the use of electrically powered reaction wheels (also called control momentum wheels). These often fairly heavy, solid metal wheels with fixed axes are used to rotate the spacecraft in different directions.

If a wheel is set in motion with an electric motor, not only will the wheel start to rotate, but the space probe itself will also begin to spin in the opposite direction. While the reaction wheel is rotating on the spacecraft, the spacecraft is rotating with respect to the wheel. This is another manifestation of Newton's "action = reaction" principle that is also behind the working of rocket motors.

FIGURE 3.12 ESA's Rosetta comet explorer, here seen during assembly, is a typical example of a three-axis stabilized spacecraft. [ESA]

In reality the wheels are always and continuously turning in one direction only, and the rotating force is obtained by acceleration and deceleration. Accelerating a wheel has the same effect as spinning it up from standstill. Deceleration is equivalent to spinning it in the opposite direction.

Since a spacecraft has three orthogonal axes around which it can rotate (one for left–right or "yaw" rotations, one for up–down or "pitch" rotations and one for "roll" rotations), it can in principle be controlled by three of these reaction wheels (although there are smart ways to do it with only two). To make the spacecraft rotate in one direction, you spin up the proper wheel in the opposite direction. To rotate the vehicle back, you slow down the wheel. A fourth wheel is usually included as a spare, in case one of the other three breaks down.

If a wheel is spun more often in one direction than in the other, after a while it will turn at its maximum velocity. Reaction wheels can typically reach speeds of 6,000 revolutions per minute! The wheel then has to be slowed down. Without any compensation, decelerating the wheel would make the spacecraft rotate, but by using opposite thrust from the reaction control system this can be avoided. Such an action is called "momentum desaturation" or "momentum unloading." Once the wheel regains a reasonable velocity, the reaction control system can be disengaged and the wheel can be again used for rotation control.

Although most spacecraft use thrusters for these momentum desaturation maneuvers, some spacecraft contain sensitive optical equipment that can be contaminated by the exhaust of the rocket thrusters. The Hubble Space Telescope, for instance, uses magnetic torquers instead of thrusters.

Magnetic torquers are essentially long electromagnets that align themselves with respect to an external magnetic field when electricity is flowing through them. The effect is similar to that of a compass needle lining itself up to the direction of the Earth's magnetic field. As you need a relatively strong magnetic field of a planet to use magnetic torquers, they are mostly used on Earth-orbiting satellites and seldom on interplanetary probes.

A strong benefit is that, unlike thrusters, they only require electrical power. That means you can do momentum desaturation maneuvers without consuming any of the precious propellant.

Three-axis-controlled spacecraft can point antennas and instruments continuously in any given direction, without having to de-spin them. However, while a probe orbits a planet, it requires a slow, continuous rotation to keep its cameras pointed at the surface. It will also need to change its attitude to keep its antenna pointed to the Earth.

If thrusters are used for routine stabilization and attitude change maneuvers, this will require a relatively high amount of propellant. Moreover, it means that the spacecraft is always slowly rocking back and forth, because there are limits to how precisely a thruster can be fired. Often you will overshoot and need another impulse to compensate, which in turn is bound to be less precise than required and therefore needs compensation itself. This can complicate long-duration observations with optical cameras that have to be pointed in precisely the same direction for extended times.

The use of reaction wheels results in a much steadier spacecraft, but they can be rather heavy, suffer from relatively high failure rates, and have limited mechanical lifetimes that need to be taken into account.

The choice for the type of stabilization the spacecraft designers make depends on what the space probe is supposed to do, how accurate its measurements have to be, and how large and heavy the design can be allowed to become. That is why each method of stabilization, spin and three-axis, with rocket thrusters, reaction wheels or otherwise, is still being used in modern space probes and satellites today.

Irrespective of how it is stabilized, if a spacecraft is to point itself in the right directions it needs to know what those directions are. On Earth, navigators on board ships and airplanes used to look at the stars, the position of the Sun and the direction of a compass needle to find out where they were going. Modern space probes employ similar methods, and carry a range of sensors that tell them where they are and what their attitude is.

"Sun sensors" indicate to a spacecraft the direction of the Sun. They come in two basic varieties: coarse and fine. "Coarse sun sensors" can only tell if the Sun is in their field of view, and are thus used only for very general direction determination. They usually consist of a simple solar cell that only gives off electricity if the Sun shines on it. Coarse sun sensors may, for instance, be useful to help to point the solar arrays at the Sun and keep onboard electricity flowing in case of a serious attitude control failure.

"Fine sun sensors" measure the Sun's direction – and therefore the space probe's attitude with respect to it – much more accurately than coarse sun sensors. A simple version may consist of a solar cell and a device that measures the amount of electricity it generates. The more directly the Sun shines on it, the higher the voltage. Others employ constructions that cast shadow patterns on solar cells. The location and shape of the shadows then indicate the precise direction of the Sun.

"Horizon sensors" are used to find a planet's horizon from an orbiting spacecraft. They usually consist of an infrared detector that measures the abrupt change in infrared radiation when it sweeps from deep space onto the planet, or vice versa.

However, they are only good for giving very rough directions, and usually depend on the radiation from carbon dioxide in a planet's atmosphere. Most of them would not work on a moon without an atmosphere, and they would also easily be confused by the radiation from a large mother planet behind a moon.

"Star trackers" are optical sensors that can be pointed at a single star. By tracking how this star is moving in the sensor's field of view, a star tracker can tell the spacecraft the direction in which it is moving or rotating. By keeping a star tracker trained at a certain star, the side of the space probe on which it is located is always pointed in the same direction.

"Star mappers" are similar to star trackers, but follow more than one

FIGURE 3.13 Star trackers like these tell a spacecraft its attitude with respect to the stars. [EADS Space]

star in their field of view. By remembering the positions of several stars with respect to each other, their electronics can determine the orientation of the spacecraft in all three dimensions.

Using a star tracker is like finding a certain constellation on a dark night, or a star in that constellation, to determine the direction you are looking. For instance, the North Pole Star, Polaris, always tells you where to find the north.

Using a star mapper is like using the shape of a constellation to determine not only your direction, but also your own relative position; if you see Orion or the Big Dipper upside down, you must be standing on your head (or you somehow have ended up on the other hemisphere).

Star trackers and mappers can normally only be used on three-axis-stabilized spacecraft or on de-spun platforms of spinning spacecraft. On a spinning spacecraft you usually need "star scanners," which keep track of stars quickly flying through their field of view to determine the spacecraft's attitude with the help of a programmed star catalog.

Magnetometers measure the direction and sometimes the strength of a local magnetic field. They are relatively simple devices, but you need a relatively strong magnetic field for them to work. Moreover, planetary magnetic fields usually have all kinds of fluctuations and abnormalities. These have to be mapped in detail for the magnetometers to give anything more than a very crude sense of direction. Magnetometers as attitude control sensors are therefore mostly used on Earth-orbiting satellites, but as scientific instruments for measuring magnetic fields they are widely employed on interplanetary spacecraft (we shall discuss this in a later chapter).

Gyroscopes do not indicate actual directions with respect to the outside world, but accurately tell a spacecraft the directions and velocities of its own rotations. Traditional gyroscopes are like spinning tops that always point in the same direction, measuring how the rest of the spacecraft is rotating around them.

Modern Ring Laser Gyros use a laser beam that is split and projected into two ring-shaped cavities, pointed in different directions. Because the laser generates a very coherent beam of light, both of the beams will have the same, very specific frequency.

When the spacecraft and the gyro are rotating, the wavelengths and thus frequencies of the normally identical laser beams diverge a little with respect to each other.

This Doppler effect is similar to the different sounds a train makes when it is coming toward you and when it is moving away. When it rushes past, you hear the frequency of its sound changing from high to low.

When the wave patterns of the two beams are combined by two opposing mirrors arranged to reflect precisely against each other, they result in a certain interference pattern. (For example, if you throw two stones into a pond, the resulting waves will meet. At certain places the result will be higher waves; at other places the "valleys" and "hills" of the waves will cancel each other out.) Due to the Doppler effect, the interference pattern of the two beams at rest is different from when they are rotating, and in a Ring Laser Gyro this principle is used to measure how rapidly a spacecraft is turning.

Starting with a known attitude, and by keeping track of the spacecraft's rotations using gyroscopes, you can still get a good idea of its new attitude after some movements. It's something similar to standing in a room in the direction of the door, and have someone turn you around with your eyes closed. As long as the movements are not too fast or too many, you will still be able to tell where the door is at any time.

Gyroscopes are often combined with accelerometers in so-called Inertial Measurement Units (IMUs). The accelerometers measure the rate of change in velocity over a specific period of time, and can therefore tell the spacecraft when it has fired its rocket engines for the sufficient length of time. The gyroscopes measure how fast the spacecraft is turning and can thus be used to control rotation speeds. Gyroscopes are also useful to estimate the spacecraft's orientation for short periods, when for example the spacecraft is turning too quickly for the star trackers to work properly. IMUs usually have three gyroscopes and three accelerometers – one combination of both for each spacecraft axis.

For accurate attitude determination and navigation – and sufficient amounts of redundancy in case equipment fails – most space probes combine the various types of sensors described here. It's no wonder that the design of the attitude control subsystem can easily becoming very complicated.

PROPULSION

When putting an interplanetary spacecraft on our imaginary dissecting table, we had better make sure that its potentially dangerous rocket propulsion system is inoperative. Interplanetary spacecraft usually have one or more relatively large rocket motors, of a size you normally do not see on Earth satellites. They need these to make trajectory adjustments, brake to get into an orbit around another planet, and change their orbit altitude and inclination (the angle with respect to the planet's equator).

FIGURE 3.14 A liquid propellant rocket engine that can be used for attitude control or for orbital maneuvers. [EADS Space]

The operation of chemical rocket engines has been explained earlier in this book (see "Rockets and Satellites" on page 4). As on launch vehicles, most rocket propulsion subsystems on board spacecraft consist of tanks with propellants, rocket engines, and pipes with various valves to transport the propellants from the tanks to the engines. The propellants are either pumped into the engines or forced into them by high-pressure gas from another tank.

Some propulsion subsystems use a single type of propellant, called a "monopropellant," which can burn on its own with the use of a chemical catalyst like platinum or iridium (a catalyst is a chemical substance that helps a chemical reaction but is not actually part of it; it's a sort of molecular helping hand). Hydrazine is a commonly used monopropellant.

More powerful spacecraft engines use bipropellants, with an oxidizer (typically nitrogen tetroxide) and a fuel (usually monomethyl hydrazine) stored in separate tanks. These are burned together in the engine, usually igniting automatically when mixed (which is very handy, but it also means that they have to be kept absolutely separate until needed).

Sometimes boosters with solid propellants are used. These are basically cylinders with a nozzle, filled with a solid body of a fast-burning substance (called the propellant "grain"), much like the rockets fired at New Year's Eve.

Boosters are powerful and relatively simple, lacking all the complicated plumbing needed for liquid propellants, but they can only be used once. After ignition, the motor just burns till the propellant has been exhausted. It cannot be throttled up or down, it cannot be stopped, and, like a match, it cannot be tested before actual use. An additional liquid propellant subsystem is therefore usually needed for the more gentle, repetitive boosts.

Solid propellant motors are used to kick spacecraft into or out of an orbit – for instance, to speed them from Earth orbit into an interplanetary trajectory. The firing of these motors stands high on the list of most nerve-wracking moments in a spacecraft's operational life. If they don't ignite, space probes get stuck in Earth orbit and will never reach the Moon or another planet. If they do, solid propellant boosters sometimes violently explode; then your expensive spacecraft is not only stuck, but also in pieces. NASA's "Contour" comet explorer recently suffered this awful fate.

If you want to use a solid propellant rocket motor later in the flight – for instance, to launch something back from Mars – you have to make sure that the long exposure to extreme temperatures in space has not affected the propellant. The grain may have been cracked, or its chemical properties may have changed so that the motor will give a lower thrust.

FIGURE 3.15 The Inertial Upper Stage is a large solid propellant rocket motor used to boost interplanetary spacecraft on their way. [NASA]

Electric propulsion subsystems such as that on SMART 1 (see the "Proving Technology" section on page 33) generally include similar equipment as conventional liquid propulsion subsystems, except that the tanks can be much smaller and the actual engines look very different.

The reaction control subsystem of a spacecraft is also usually considered to be a part of the propulsion subsystem, often feeding off the same tanks as the main rocket engine. Reaction control thrusters work in a similar way as the large main rocket motors, but produce much lower thrusts. They are used to rotate the spacecraft around its own axes, and are thus a means of controlling its attitude.

Simple subsystems on some small spacecraft use cold gas thrusters. Just like a party balloon, these thrusters give a push simply by opening a valve and letting out a pressurized gas (usually nitrogen) without the need for any combustion. These are relatively cheap, but are not very powerful.

Small electric propulsion thrusters, usually fed with Xenon gas, can also be used for attitude control, but their thrust is even weaker than that of cold gas thrusters. Nevertheless, their high propellant efficiency makes them very interesting for long-duration missions that only need gentle maneuvers with little attitude control thrust. In such cases they can save a lot of propellant tank volume.

The small thrusters for attitude control are sometimes also used for orbit maneuvering. In such cases the thrusters are used in pairs on both sides of the spacecraft, to ensure that they do not induce rotations but change the spacecraft's velocity in a certain direction.

DATA HANDLING

Traditionally, humans act as the brain of such complicated moving machines as cars, jet fighters, cranes, tanks, submarines, cruise ships and helicopters. Unmanned military observation airplanes flying over dangerous areas are often remotely controlled by people on the ground.

However, this can only work if the operator is on board the vehicle or is at least close enough to control the system remotely without long delays. Space probes are often too far away to be directly controlled by radio as the signals may take hours to travel from the control station on Earth to the spacecraft and back.

Moreover, modern spacecraft are often far too complex and things happen too quickly for a human to be in direct, real-time control. That is why most space probes really need to be robotic.

Today's space probes can have several computers on board, but one main computer is usually responsible for the overall management of the activities; the brain of the spacecraft.

This computer maintains the timing and the sequence of what needs to be done, interprets commands from Earth, and collects, processes, and formats the telemetry data that needs to be sent to Earth. It may also interpret the signals it receives from the star trackers and other sensors and activate reaction control thrusters or reaction wheels to correct the attitude, although that is often the job of a separate attitude control computer.

In the old days, programs were often "hard-wired"; the software was a fixed, permanent part of the computer and could not be changed after launch. Nowadays software can be improved and adapted to new situations, and be uploaded to the spacecraft by radio to overwrite previous programs. Space probes are sometimes even launched before all their onboard software is ready. They start out with only the basic operating instructions needed for the first part of the mission, while the rest is being developed for uploading later.

Modern spacecraft computers are based on commercial chips that are also used in home computers, but are significantly modified to ensure that they survive the much higher radiation levels in space. The computer of NASA's latest Mars explorer, the Mars Reconnaissance Orbiter, for instance, employs a new generation of space-qualified processors, based on the 133-MHz PowerPC processor. The computer you have at home is probably much faster and working at Gigahertz speeds. However, your PC would not last very long above the Earth's atmosphere, with all the radiation and extreme temperatures there.

Spacecraft working far from Earth and doing difficult work have to be especially autonomous. For example, radio signals from the Cassini probe, currently in orbit around Saturn, take one and half hours to reach our ground stations. A command from us to the spacecraft then takes the same period of time to reach Cassini. If the spacecraft had no autonomy, manual operation with such three-hour delays would be extremely difficult and make everything work in slow motion!

Moreover, viewed from Earth, interplanetary spacecraft may at certain times orbit behind the Sun for over a week, or for a shorter time move behind a planet. When that happens communication is impossible, and the spacecraft is required to survive on its own for the duration of that phase. Crucial orbital maneuvers involving rocket firings often take place while spacecraft are out of communication, so a certain level of autonomy is absolutely necessary.

The spacecraft computer also controls the scientific instruments and stores the data they collect. If there are lots of scientific instruments to handle, a separate Payload computer is sometimes installed.

To store data from the instruments, modern space probes use Solid State Recorders. The one on Mars Reconnaissance Orbiter (MRO) has a total capacity of 160 Gigabits, which means it can hold the data equivalent of 40,000 pop songs. Instead of music, science data is stored until it can be transmitted to Earth, and then overwritten with new data. A capacity of 160 Gigabits may seem like a lot, but it may require up to 28 Gigabits to store a single image taken with MRO's HiRISE instrument system!

Solid State Recorders are so called because, unlike tape recorders and hard disk drives, they have no moving parts. Instead they are based on several hundreds of memory chips for storing digital data. Without moving elements, they are much more reliable and less likely to break down during many years of continuous usage.

The first interplanetary spacecraft to use Solid State Recorders instead of tape recorders was the Saturn orbiter Cassini, which reached the ringed planet on July 2004. The previous large interplanetary NASA mission, the Galileo orbiter that was sent to Jupiter, had to fly with old-fashioned tape recorders, and they caused a lot of problems.

In the past, all the measurements of all the onboard scientific instruments were recorded and sent to Earth via radio. Separating the important information from the redundant data and useless noise was done on Earth, with large and heavy computers requiring lots of calculation time.

However, improvements in spacecraft technology are often accompanied with increases in the amount of data that needs to be stored and transmitted. The latest instruments often produce far too much data to send over the communications link to Earth; it would take too long and require excessively large transmitters that would use too much electrical power. Fortunately, also in this respect, spacecraft are getting smarter. Modern computer technology allows much of the raw data to be sorted out on board the spacecraft, and only the much reduced, partly analyzed data needs to be sent back to Earth for further examination.

The onboard computer relies on complicated software programs to run everything on board. The software aims the solar arrays and antennas, directs temperature regulation equipment, prepares data for transmitting to Earth and puts scientific instruments on and off.

The onboard software also tells the spacecraft what to do when something goes wrong – for instance, if it loses track of its attitude or its critical equipment is malfunctioning. It then makes the probe go into "safe-mode," and the robot puts itself into a stable situation to await orders from Earth. It's like a child who has lost his or her parents in a busy supermarket, just waiting around at the same place until one of them shows up.

In safe-mode a spacecraft only operates at a minimum, not doing much other than directing its solar arrays at the Sun and its main antenna at the Earth to send housekeeping telemetry (data on the functioning of its subsystems) back to its operators. Safe-mode is basically a sort of standby function on the spacecraft.

If something goes seriously wrong with the attitude control system, the probe may completely lose its sense of direction. It then no longer knows where the Sun is, where the Earth is, or what its attitude is. In this case it goes into "survival mode," operating at the truly bare minimum. Typically the spacecraft will then send out simple distress signals (the child in the supermarket example is now also screaming) and use its coarse Sun sensors to roughly find out where the Sun is. It will also start to rotate in the hope that the Earth will eventually move into view of the antennas so that the ground stations can receive the calls for help. The ground controllers can then tell the probe were to point its antennas, and start to receive more detailed telemetry data to help to find solutions to the problems on board.

As space robots get more and more autonomous, they are increasingly capable of handling difficulties themselves, without intervention by ground control support. Often they can even predict that equipment is about to malfunction – for instance, if the temperature of a reaction wheel is getting too high due to increased friction. They can basically "feel," just as we would put down heavy shopping bags when our arms started to ache, to prevent us from pulling a muscle.

An interesting example of an important piece of onboard software is the Command-Loss Timer. Every time the spacecraft is in radio contact with Earth this timer is reset to a certain value, for instance a week. When the spacecraft loses contact with its controllers on the ground, the timer starts to count down.

If it reaches zero – meaning that for a week it hasn't heard anything from control – it starts to operate following the pre-programmed assumption that its receiver or something else is broken (just as you might check whether your phone is still working if your mother hasn't called for some time). Other software then commands the spacecraft to switch to backup equipment such as the redundant receiver, to see if it can get a signal that way.

The spacecraft may also start to sweep the sky with its antennas to try to find the radio beam from Earth, just in case there is something wrong with its attitude determination and control subsystem.

Once the probe is in contact with Earth once more, the ground control teams can figure out what happened. Sometimes they can correct the problem by sending updated software to the spacecraft and remotely installing it on its computer. They can also permanently deactivate faulty equipment and tell the spacecraft never to use that faulty part again.

MECHANISMS

The mechanisms on a spacecraft have the same function as the joints in our bodies; they are used to rotate, deploy, eject, open and close things. Designers always attempt to keep the number of mechanisms in a spacecraft as low as possible, because they have a relatively high chance of failure and usually a breakdown is critical to the mission.

For instance, if a mechanism that has to release a folded solar array does not work, the spacecraft will not be able to generate enough electricity or even, perhaps, no power at all. Galileo's stuck umbrella antenna mentioned before is another example of how a problem with one relatively simple mechanism can severely upset mission plans.

Developing mechanisms for space is pretty hard, mainly because of the high temperature changes they have to endure and the need to maximize their reliability. Lubrications used in mechanisms on Earth, such as oil or grease, evaporate or freeze in space. Moreover, if a space probe mechanism gets stuck in space there is no one around to give it a kick and get it working again.

Mechanisms for the deployment of solar arrays, antennas etc. are usually activated soon after launch, to give the space environment as little chance as possible to have adverse effects. Deployment mechanisms are normally needed only once; after they have done their work it is no longer important whether their lubrication dries up or their wheels freeze and stick.

For these one-shot mechanisms spacecraft engineers often employ pyrotechnical devices – small explosives set off by a small spike of electric current. These are small, simple, cheap and relatively reliable, but they can be used only once. As the explosives can be dangerous to people, much care for safety has to be taken during testing and integration with the spacecraft on the ground.

Pyro devices are used for various tasks: to separate a spacecraft from its launch vehicle, to permanently deploy antenna booms; to release instrument covers, to permanently open or close valves in propellant lines; to jettison heatshields; and to deploy parachutes on lander systems.

NASA's Mars Pathfinder mission with its Sojourner rover depended on the operation of 42 pyro device activations during entry into the Martian atmosphere, the descent and finally the landing. First the 11-meter (36-foot) parachute had to be shot out of the lander. Then the front heatshield was pyrotechnically ejected, followed by the separation of the lander from the protective backshell.

Just before hitting the ground, the airbags were released by pyrotechnically cutting the ties on their wrapping during the flight to Mars. These airbags were subsequently inflated with gas generated by burning small amounts of solid propellant in three separate gas generators.

Solid propellant rocket engines were fired to slow the lander down in the final seconds before landing, and the lander was finally cut free from the parachute with a pyro device. Like a giant beach ball Pathfinder then bounced over the surface of Mars 15 times for almost 1 kilometer (0.6 mile), rising as high as 15 meters (50 feet).

More recently, each of the larger Mars Exploration Rovers, Spirit and Opportunity, fired no less than 126 pyro devices during the same mission phases.

Other mechanisms need to work properly and continuously during the entire mission, such as antenna-pointing mechanisms and the Solar Array Drive Mechanisms that keep the solar arrays aimed at the Sun. These mechanisms require special attention and are therefore rather expensive, in the order of half a million dollars each.

Special mechanisms are those on planetary rovers, such as the electric motors driving the wheels. Their lifetime is not only measured in absolute time, but also in revolutions – that is, the number of times the wheels have gone round. There is a lot of wear and tear by Moon or Mars dust that gets into delicate mechanisms and grinds carefully polished surfaces each time a wheel makes a turn. Given enough time, the wheel mechanisms will be so badly damaged that the internal friction becomes higher than the engines can overcome, at which moment the rover will be stuck. The lifetimes of planetary orbiters are measured in years, but the birthdays of a rover on the surface of Mars are celebrated a month at a time.

On the other hand, knowing what may happen to spacecraft equipment used in harsh environments may inspire engineers to make their mechanisms so sturdy that they actually perform much better and longer than strictly required. The NASA rovers Spirit and Opportunity were, for instance, designed to last three months on the surface of Mars, but at the time of writing they are still going strong over two years after landing.

THE KIDNAP OF LUNA 1

As we are curious about how a modern interplanetary spacecraft works, in our imagination we have put one on an operating table and investigated it inside and out. However, in a *Spaceflight* magazine article called "Those magnificent spooks and their spying machines," author Dwayne A. Day describes how the CIA once did such a thing for real!

In 1959 the Cold War was running hot in space. The Soviets were launching a series of ever-larger satellites on their big launchers, and the USA was scrambling to keep up. In January of that year the Soviets even managed to pass within 6,000 kilometers (3,700 miles) of the lunar surface with their Luna 1 spacecraft, although they lost contact with it after 62 hours. America felt it was seriously behind in space technology and really wanted to get a good look at its rival's spacecraft.

At the time, both superpowers were trying hard to convince the rest of the world that their social economic systems were the best. The Soviets and Americans were especially struggling for influence in the Third World countries, where many governments had not yet made a choice which of the two they would prefer to act as their big brother.

Space achievements were valuable propaganda material for showcasing technological superiority, and a real production model of the Luna 1 probe was displayed at the sable and mink pelt exhibition in Mexico City. The spacecraft came complete with its rocket insertion stage and covering launcher payload fairing, in which three windows had been cut to enable the public to view the inside of the small lunar probe.

After the show, the spacecraft and accompanying displays were to be carried by truck to a railroad station for further transport to Veracruz by train. The Soviets had placed an inspector at the station to check and log the arrival of each truck. However, they had not given him any means of communication with the fairground from which the trucks were departing. Moreover, the trucks drove without any escort.

The CIA saw an amazing opportunity and its agents arranged for the Luna spacecraft to be on the last outbound truck of the day. They stopped it at a turnoff and drove it to a salvage yard they had rented.

CIA experts spent the evening and most of the night thoroughly inspecting Luna 1 and gaining valuable knowledge about the secrets of Soviet space technology. In particular, they learned which Soviet companies were the producers of the spacecraft and its components, and the kind of guidance system that Luna 1 used. In addition, they found out that this particular model was the fifth of its type that the Soviets had constructed.

About an hour after the experts left, a CIA driver drove the truck to a pre-arranged point to meet the original driver, who then took over the vehicle and continued his journey to the train station. The Soviet inspector arrived shortly afterwards, logged the crate with the spacecraft, and had it loaded on a railroad flatcar. The train left with all the crates, and the Soviets apparently had no suspicion of what had happened between that morning and the evening before.

Sometimes spacecraft engineering can be as glamorous as a James Bond movie.

4

BUILDING AND TESTING

NOW, clear your mind from the mess we created by dissecting an imaginary spacecraft, and let's have a look how it really works – how spacecraft are actually built and tested.

THE SPACECRAFT GARAGE

Spacecraft Prime Contractors, the companies responsible for the overall design and construction of spacecraft, assemble them from the loose pieces of equipment described in the previous chapter. This equipment has been designed, built and tested separately, often by other companies acting as subcontractors in the project.

Like a garage full of car wheels, doors, windows, bumpers and engine parts, each of the equipment units has been individually checked for quality and should fit perfectly into the final product. However, the big challenge remaining is to put it all together into a working machine; good parts do not necessarily make a good car.

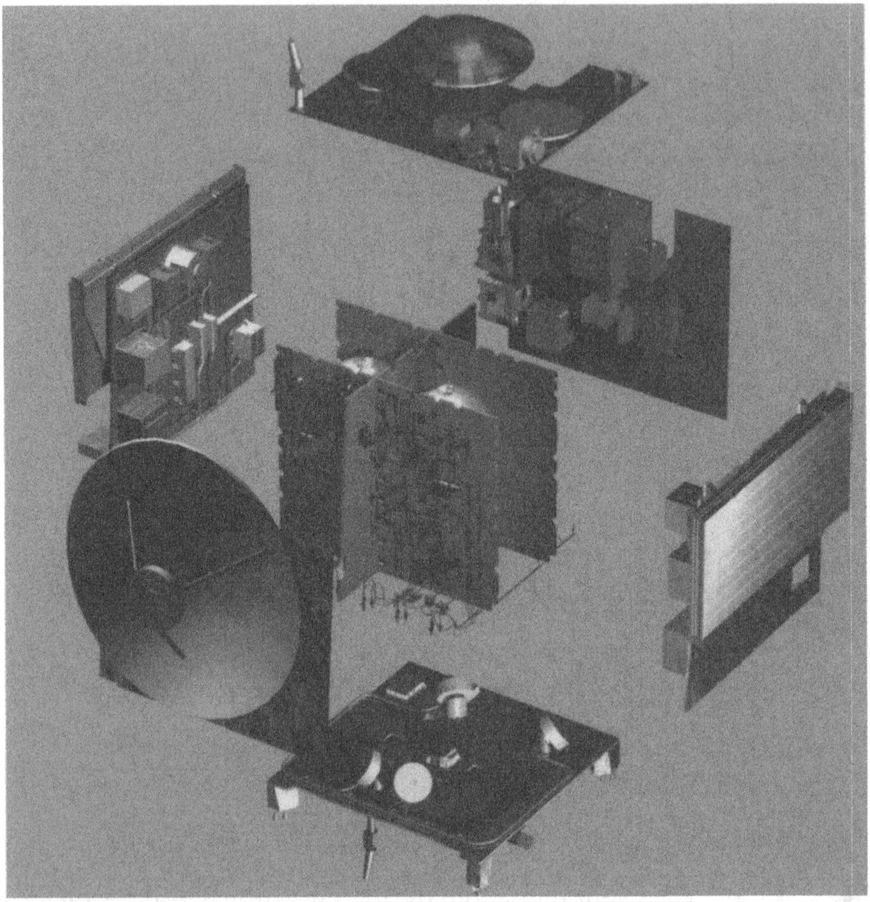

FIGURE 4.1 An exploded view of Mars Express, showing the various equipment units and the layout of the spacecraft structure. [ESA]

How their equipment connects with the rest of the satellite is a special concern to equipment designers. All the equipment built by all the different companies has to match perfectly into a single spacecraft. Often this cannot be tested until all the equipment is ready and the assembly of the spacecraft has actually been started.

Building interplanetary space probes requires great care, as the equipment from which a spacecraft is assembled is very expensive, sometimes fragile and often unique. A falling tool can easily damage the lens of a multi-million-dollar telescope instrument and delay a mission by months.

In 1968, a technician working on an astronomical observatory satellite named "Copernicus" tripped over a piece of scrap lumber while helping to carry the main mirror for the spacecraft's telescope. The accident destroyed the million-dollar mirror and seriously threatened the spacecraft's production schedule, because it took nine months to build a new one.

In 2004 an almost finished NOAA meteorological satellite fell off the ground support trolley to which it was supposed to be securely bolted. The spacecraft smacked onto the hard floor, resulting in millions of dollars of damage. The bolts were missing: someone forgot to put them back after a test, and another person forgot to check if they were really there before moving the trolley with the spacecraft on it.

If parts of a spacecraft break, you cannot go back to the shop to get a new one in time for the launch, and nor can you make some quick repairs. Taking a spacecraft apart, manufacturing replacement parts, reassembling and retesting can take many months.

Moreover, undiscovered assembly mistakes cannot be solved *after* launch, so great care is taken over even the simplest action. Typically one engineer is working on something while another is checking everything the other person does, all the time maintaining a list of all the actions performed. Building a spacecraft is therefore a slow and meticulous process.

FIGURE 4.2 A small army of technicians is preparing the Messenger spacecraft for a vibration test, to see whether it is able to withstand the rough ride on a rocket. [NASA]

Chapter 4

KEEP IT CLEAN

Spacecraft are built and tested in so-called "cleanrooms" that are cleaner than a normal hospital operating theatre. A cleanroom is an area where the air is filtered to contain only a very limited amount of dust particles. Dust can blind sensitive optical instruments and sensors. It can also mess up the measurements of equipment that determines the composition of dust particles in comets or on other planets. It is essential to make sure that what is being measured is really space dust, and not Earth dust that has been caught inside the instrument!

A cleanroom can never be completely dust-free, but it can be filtered to various high levels of cleanliness. Most spacecraft require so-called Class 100,000 cleanrooms for construction and testing, meaning that there must be fewer than 100,000 dust particles larger than 0.5 micrometer (1.6 millionth of a foot) per cubic foot (about 27 liters) of air. To give you an idea of how clean this is: the room in which you are reading this right now probably contains several million dust particles per liter of air! The Beagle 2 Marslander was even constructed in a Class 100 facility – an area with less than 100 dust particles over 0.5 micrometer per cubic foot.

A major source of dust in a cleanroom are the people working in it. That is why, before entering a cleanroom, you have to put on a dust-free "bunny suit": a long lab coat, gloves and elasticized covers over your head and shoes. Then you walk across a sticky floor-mat to remove dirt from your soles. Cleanrooms of Class 10,000 and better also require people to wear surgical facemasks.

Humidity and temperature also need to be controlled. Humidity must be kept fairly low, typically within a range of 40 to 55 per cent, i.e. the air should contain no less than 40 per cent and no more than 55 per cent of the maximum amount of water vapor it can carry. The temperature inside a cleanroom is usually about 20 degrees Celsius (68 degrees Fahrenheit).

The need for cleanrooms goes together with the increased sensitivity and therefore fragility of spacecraft hardware. The Russians, with their more ruggedly designed equipment, are known to assemble their spacecraft in factory-like facilities. Discarded tools and rusting pieces of scrap are lying around in their assembly halls and launch preparation buildings. Nevertheless, their spacecraft generally work well.

Small satellites built by universities and radio amateurs often do not follow the conventional cleanroom approaches either. Apparently, cleanrooms of a high class are not always needed, but to have a mission worth hundreds of millions of dollars fail because of some dirt on a critical detector is an unacceptable risk.

FIGURE 4.3 Cleanrooms can be very large. Here the Rosetta comet lander is hoisted on top of its mother spacecraft in ESA's test area. [ESA]

Chapter 4

PLANETARY PROTECTION

Assembling equipment and spacecraft becomes even harder if "Planetary Protection" measures have to be taken. These are implemented to avoid the pollution of planets with Earth microbes, and to avoid the Earth from being contaminated by microbes from space. Although this sounds like something hypothetical from a science fiction movie (think of the Man in Black "protecting the Earth from the scum of the Universe," the sick alien invaders of *War of the Worlds*, or *The Andromeda Strain*) it is in fact a very serious business.

Most places in the Solar System are regarded as completely dead. Life as we know it could never survive the heat and harsh radiation on Mercury, or the lack of water and air on the Moon. However, some unknown microbes or simple plants may still exist on Mars or live deep under the surface of some moons of Jupiter. William Randolph Hearst, the famous American publisher, once sent a telegram to a leading astronomer asking "Is there life on Mars? Please cable 1,000 words." The astronomer replied with "Nobody knows," repeated 500 times.

We must make sure that our space probes that visit those places do not accidentally contaminate the environments of these possible indigenous lifeforms. Some especially tough Earthly microbes might be able to survive on Mars. Multiplying rapidly, they could contaminate the planet and forever change the environments in which genuine Martian microbes may be living.

An analogy often used is that of the rabbits that early colonists brought to Australia. Some escaped from the farms, and without sufficient numbers of natural enemies they multiplied rapidly and soon turned into a plague.

Another historic example is that of the rats that commonly lived on the old sailing ships. When these ships visited deserted islands to look for water and food, the rats sometimes managed to leave the vessel. Unable to protect themselves from the suddenly appearing strangers, flora and fauna on many formerly pristine islands was totally ruined. The death of many Native Americans due to diseases brought to America by European explorers and conquerors is another lesson from history.

The problem could also happen the other way around: the Earth environment could potentially be contaminated by Martian microbes carried on board future spacecraft bringing back samples of the Martian soil and atmosphere.

When the Apollo 11 crew returned from their historic voyage to the Moon, they were immediately put inside a hermetically sealed caravan.

FIGURE 4.4 The Apollo 11 astronauts greeted by president Nixon, while being locked up in their quarantine caravan. [NASA]

They had to stay there for three days, until the doctors were reasonably certain they had not been infected by possible "Moon germs." After later Apollo lunar landings confirmed that nothing lived on the Moon, these somewhat unfriendly reception procedures were no longer applied to Moon missions.

The moonrocks the moonwalkers brought with them were handled and examined in a specially built facility, to avoid moon dust or lunar microbes getting into the Earth's environment. These lunar samples are still stored in this place, to keep them absolutely clean and pristine. However, no one fears a possible "lunar plague" nowadays.

In contrast, material from Mars is still considered to be potentially dangerous, and when future spacecraft eventually bring some to Earth it

FIGURE 4.5 Moonrocks like this one collected by the crew of Apollo 15 were once feared to contain lunar microbes that might contaminate the Earth. [NASA]

will have to be treated with the utmost care. Samples scooped from the surface or drilled up from below it must be hermetically sealed. No surface of the spacecraft that has been exposed to the Martian ground or atmosphere will be allowed to come in contact with the Earth environment upon return. This results in very complicated packaging mechanisms and difficult sample transfers between the spacecraft launched from Mars and the separate spacecraft flying back to Earth.

Moreover, we must be absolutely sure that when the container with Mars samples returns to Earth it will in no circumstances burn up during re-entry or crack open when landing. As parachutes can fail, the return capsule cannot rely on them and has to be designed in such a way that it can survive re-entry and a high-velocity impact without any active stabilization or braking mechanisms.

The NASA and ESA designs for such capsules pack the sample containers in thick layers of foamy material inside a thick self-stabilizing heatshield. The capsules will not have parachutes, because they must be able to land without them. Future Martian rocks brought to Earth will arrive with a bang!

Cleaning spacecraft equipment and keeping it sterile during assembly and testing is very difficult, slows down the spacecraft construction and is rather expensive. It requires especially clean facilities, assembly procedures that minimize the amount the spacecraft needs to be touched, and sterilization procedures with dry heat, gamma radiation, gas plasma or disinfecting wipes.

The fully assembled spacecraft needs to be sealed inside an airtight "bio-shield" and is sometimes heated to 120 degrees Celsius (250 degrees Fahrenheit) for final sterilization. Putting the spacecraft in an oven kills most microbes, but can also destroy some materials normally used in the spacecraft, like certain plastics and electronics. Specially developed high-temperature equipment may be needed, thus increasing costs and complexity even more.

Working on spacecraft with severe Planetary Protection requirements is a very slow and difficult process. Engineers have to wear chirurgical outfits while working on the spacecraft, and their gloves make it difficult to manipulate small parts.

The assembly room for the Beagle 2 Marslander even had a large glass window behind which a Planetary Protection expert continually watched the assembly technicians work. He kept a list of how many times a part of the lander was touched, ordering cleaning measures when a certain limit was reached. He also checked whether the engineers were not accidentally breathing on the spacecraft through their surgical masks.

If they put their hands in their sides too long (a habit of spacecraft assemblers because it prevents them from accidentally touching things), skin cells could be rubbed through their coats onto their gloves, and they had therefore to wash their hands before continuing.

Apart from seriously slowing down the assembly work, the strict supervision and continual cleaning actions also put a severe psychological load on the engineers.

Planetary Protection measures are the cause of many heated debates between those who think it is absolutely necessary, and those who regard it as a waste of time and money.

Some are convinced that the likeliness of a microbe from Earth surviving on Mars is extremely small. Moreover, they argue that possible life on Mars would in any case be so different from that on our planet that one could never influence or infect the other. Dr Zubrin, famous promoter of the Mars Direct plan for crewed missions to the red planet, likens it to the chance of our being infected by a sick tree, or vice versa.

Before Planetary Protection became a serious consideration, the Russians landed or crashed various spacecraft on Mars that were not sterilized in any way. Microbes from Earth have thus already arrived on the red planet. If these somehow felt at home in their new environment, they could have multiplied exponentially and it will already be too late for any measures to prevent them from taking over the planet.

Apart from that, impacts of large asteroids and comets have kicked rocks from Earth into space, and some must have eventually reached Mars. Likewise, we have found meteorites on Earth that, we are sure, originated from Mars, because the chemical composition of the rock and the tiny gas bubbles within is extremely similar to the composition of the rocks and atmosphere our robotic landers have found on Mars.

Microbes may have survived the trip inside these rocks, as research on the meteorites from Mars seems to point out that the inside of those rocks never reached temperatures high enough for complete sterilization. Bacteria and spores may thus have been traveling between Earth and Mars for hundreds of millions of years. If so, both planets have long since been infected with each other's microscopic life forms and our drastic Planetary Protection measures would seem rather pointless.

In 1984 scientists found a Martian meteorite (numbered ALH84001) on the Allan Hills ice field in Antarctica. The stone was apparently blasted into space some 16 million years ago, but the material inside even appeared to be about 4.6 billion years old – around the time when the Solar System was formed.

Much later, investigators found what appeared to be a fossil of a tiny microbe inside the meteorite. They also detected substances that looked like products of microbial metabolism.

In August 1996 NASA staged a big media event to announce this spectacular discovery. The story of the alien fossil was big news all over the world, and the topic of many articles and television programs. However, since then new research has cast considerable doubt on the possible alien microfossil. The debate about whether it really once was a microbe or the shape is just the result of inorganic chemical processes still rages. Also, the meteorite could have been invaded by Earth microbes after landing at the South Pole, which could explain the presence of organic molecules inside the rock. ("Organic" in this case simply means that the molecules contain carbon; all known forms of life are based on molecules with chains of carbon atoms, but organic molecules are not necessarily created by life.) Meteorite ALH84001 does not seem to offer any conclusive evidence about possible Martian life and the chances of microbes hitch-hiking to other planets.

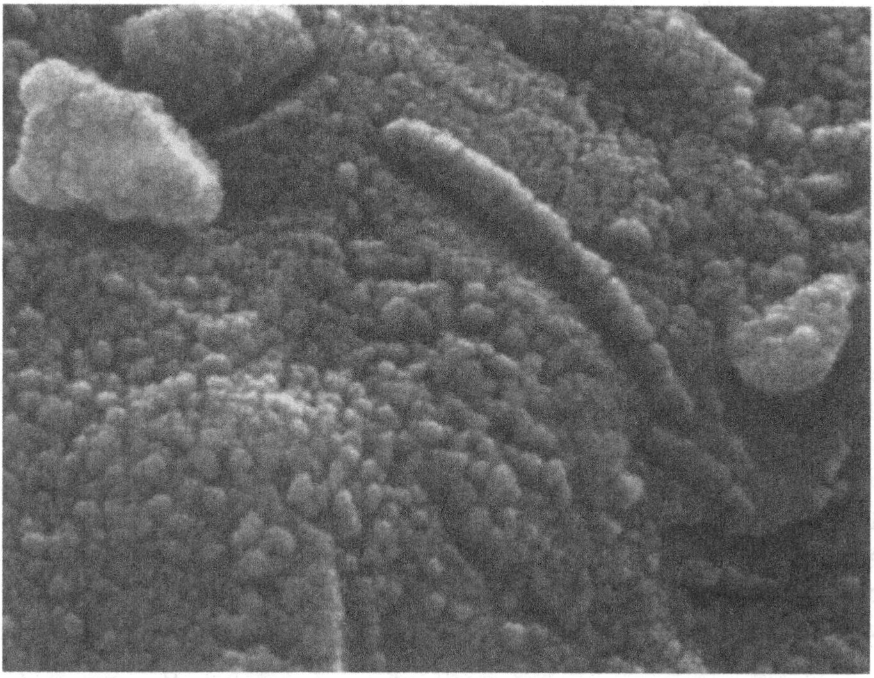

FIGURE 4.6 This worm-shaped object, less than 1/100th the width of a human hair, was found inside a meteorite that came from Mars. It might be a fossilized microbe. [NASA]

The general consensus among scientists is that we should not take chances with the potentially disastrous effects of contamination with strange, new microbes. They have made very strict, internationally agreed rules about which Planetary Protection measures are applicable to what type of missions. For now, spacecraft developers will just have to live with them.

Apart from contamination prevention, spacecraft looking for life on other planets such as Mars need to be decontaminated for another reason. If the lander were to carry a collection of microbes from Earth, its instruments could detect these instead of Martian microbes. With a "dirty" spacecraft you can never be sure that what you have detected is really from the planet you are exploring.

Sterilization measures for Marslanders looking for life are therefore more severe than those for spacecraft merely investigating the geology of the planet.

The rover on board the future ESA ExoMars spacecraft will drill below the surface and deposit samples into an automatic, onboard mini-laboratory. Inside, the material will be warmed up to see what kind of gasses come out. Some of these gasses may indicate the presence of microbes in the soil. The interior of this little laboratory must be absolutely pristine before the Mars soil is inserted. Fortunately, the oven used to warm up the samples can also be cranked up to heat itself to sterilizing high temperatures. It is in fact a self-cleaning instrument.

Shake, Rattle 'n' Roll

Because interplanetary robotic missions are expensive and challenging, we want to ensure that our spacecraft can handle the harsh experiences of launch and flying through space before we send them on their way. Just as ships are checked to be seaworthy, spacecraft are tested to be spaceworthy.

Testing of the complete spacecraft is very important, because although all the equipment individually may be functioning correctly, this does not guarantee that the whole spacecraft works as planned when everything has been put together. Often the interactions between otherwise perfectly working parts is where things go wrong.

NASA's Mars Polar Lander, launched in 1999, was designed to softly land on the south pole of Mars and investigate the terrain there. However,

FIGURE 4.7 The Mars Polar Lander, which was unfortunately lost just before landing due to a software error combined with insufficient testing. [NASA]

while descending to the surface on the thrust of its rocket engines something went wrong and once more a Mars probe disappeared without a trace. (The long list of failed Mars missions has space engineers jokingly theorize about a "Great Galactic Ghoul," a monster ferociously devouring space probes on their way to the red planet.)

Investigations to find out what had happened finally concluded that faulty software in combination with inadequate testing led to the loss of the spacecraft.

The three landing legs of Mars Polar Lander contain small microswitches that are triggered when the legs touch the surface, signaling the landing engines to cease firing. Post-accident tests showed that when the retracted legs are unfolded during the descent, the shock of their deployment accidentally triggers these microswitches. This makes the onboard computer believe that it has already reached the Martian surface and thus needs to shut off the engines. However, when the legs are unfolded the lander is actually still far above the ground. Without the rocket engines, the uncontrolled Mars Polar Lander fell down rapidly and crashed.

A couple of lines of code could have commanded the computer to ignore the landing signal during leg deployment, but it was never programmed into the lander's software. The bug did not show up during testing, because the microswitches and the deployment of the legs were tested separately. A combined test would have shown the problem, but the very tight budget did not allow for that.

The failure forced NASA to rethink its then paradigm "Faster, Better, Cheaper" approach, with its very short development phases and minimal budgets. The agency decided to allocate more money to the management and testing activities for its future space probes.

Apart from the actual spacecraft to be launched, special full-scale development models of the space probe are often built and tested during the development phase. Typical ones are Structure Models, Structure/Thermal Models, Engineering Models, and Qualification Models.

A Structure Model is used to test the spacecraft structure, mostly to see whether it can withstand a rocket launch. For missions such as Messenger and BepiColombo the Structure Model is enhanced into a Structure/Thermal Model, to test how the spacecraft will endure the severe temperatures near Mercury. An Engineering Model looks a lot like the actual spacecraft, but is built at a lower quality and without backup equipment. It is mostly used to test the electrical functioning of the spacecraft.

A Qualification Model is built to the same quality standards as the eventual spacecraft Flight Model, but is tested more severely than the actual flying probe and does not necessarily include all the backup equipment. Once the boundaries of the design's resilience have been verified, the Flight Model that will actually be launched only needs to be tested to the less demanding Acceptance Test levels. Often a Qualification Model is not needed, and the Qualification Tests are done on the finally built spacecraft. This is then called a Proto Flight Model.

Important cost savings can also be made by using the Structure Model as the basis of the Flight or Proto Flight Model, but only if it has not been damaged during testing of course.

Special facilities have been developed to simulate a launch and the vacuum and the harsh thermal and radiation environments beyond Earth's atmosphere. In these test centers we can check whether our spacecraft and its development models behave as we designed them to do. If anything does not meet our expectations, we can still correct the design of the space probe to make sure that we do not experience the same problems during the real flight.

Shaken, not stirred

Some of the first test facilities to which a spacecraft is subjected are the "shakers," mechanical vibration platforms that simulate the onslaught of a launch. Spacecraft are shaken from left to right, back and forth, and up and down; one movement at a time in most test facilities, but modern systems can shake space probes realistically in all directions simultaneously.

To avoid the shakers shaking the building as well as the spacecraft, they are built on their own independent and separate foundations, and suspended on air springs. A vibration test normally lasts only a few minutes, during which you can hardly see the spacecraft move. Nevertheless, the forces involved are severe and quite capable of destroying a badly designed spacecraft; optical lenses can be shattered in an instant, and metal panels and housings can be cracked if the vibrations are not sufficiently dampened and loads are not equally distributed over the spacecraft's structure.

Designers must make sure that the frequencies at which a spacecraft "likes" to move, the so-called "natural frequencies" or "Eigenfrequencies," are sufficiently far removed from the frequencies at which the launcher makes it vibrate. It's similar to jumping on a trampoline, where you can jump higher and higher if you find the right rhythm. Spacecraft have to be designed in such a way that their natural vibration rhythm is not the same as the rhythm of the launcher, otherwise the spacecraft may break up.

FIGURE 4.8 ESA's XMM space telescope inside the giant Large European Acoustic Facility, where the acoustic violence of a rocket launch can be simulated. [ESA]

Vibration tests can be so punishing that engineers sometimes prefer to build and test a Structural Model of the spacecraft before committing the real machine to the ordeal. If the model breaks, at least the final spacecraft can still be redesigned and the cost impact of the failure is limited.

Related to the shaker tests are the noise vibration tests, during which engineers measure whether a spacecraft can withstand the terrible noise of a rocket launch. This is tested by blasting it with sound from huge speakers inside a special chamber. The Large European Acoustic Facility of ESA's ESTEC Test Centre is a room 11 meters (36 feet) wide, 9 meters (30 feet) deep and nearly 17 meter (56 feet) high (see Figure 4.8). It has solid concrete walls of 0.5 meter (1.6 feet) thick to contain the sound.

The noise level inside the payload fairing on top of a rocket can rise to a deafening 150 decibels or more, comparable to several Boeing 747s taking off at a distance of 30 meters (100 feet). The lowest frequencies in the noise of a launch are the main problem, as they can make a spacecraft vibrate, just as the bass from a large loudspeaker can make your breastbone resonate.

Canned space

Temperatures on the outside of a spacecraft in orbit near Earth can vary from some 120 degrees Celsius (250 degrees Fahrenheit) to −160 degrees Celsius (−260 degrees Fahrenheit). Add to that the vacuum and radiation conditions, and it is clear that space is a very harsh environment.

To test how well a new space probe can handle these conditions, sophisticated space simulators are used. The big ones are large enough to contain an entire spacecraft; ESA's Large Space Simulator has a main chamber that is 15 meters (49 feet) high and 10 meters (33 feet) in diameter.

By pumping liquid nitrogen through its walls, this test chamber can be cooled to −196 degrees Celsius (−321 degrees Fahrenheit). This not only helps to simulate the coldness of space, but also to achieve a very high level of vacuum; once most of the air has been pumped out, the remaining molecules freeze to the walls of the chamber and are removed from the test environment.

The Space Simulator can also simulate the radiation from the Sun with the help of 19 Xenon lamps of 25 kW each (an IMAX theater uses just one of these lamps to project movies onto its giant screen). The lamps simulate the composition and the unfiltered intensity of sunlight above the atmosphere, and can raise a spacecraft's temperature to higher than the boiling point of water.

Inside the facility the satellite can be mounted on a special carrier mechanism that can rotate to simulate the movement of the spacecraft in orbit. The only factor a space simulator facility cannot replicate is weightlessness.

Antenna testing

Satellites and interplanetary space probes rely on antennas to communicate with the ground. These are very critical to the success of a mission and thus need to be tested.

Antenna test rooms are covered with radio frequency radiation absorbing pyramids to avoid reflections of the radio signals, as there are no walls in space. The foamy material also absorbs sound very well, like snow; people's voices have a complete lack of echo inside such a room, resulting in rather eerie, surreal sounds.

To avoid interference from television or outside radio signals, the walls of the chamber also form a steel Faraday cage that prevents any electromagnetic radiation from entering.

Wave fronts of radio signals emitted from an antenna close by are curved, like the waves in a little pond when you throw a stone into it. However, wave fronts from spacecraft in deep space arrive at the Earth as straight, parallel lines, like waves from the open sea arriving at the beach.

Radio test chambers for spacecraft therefore include special reflector panels. These bounce the curved radio wave fronts from the nearby antenna and straighten them. In this way we can simulate radio signals coming from a great distance within a relatively small test chamber.

Special cases

Some spacecraft require special tests that are not standard. For instance, they may have electromagnetic instruments on board that are sensitive to interference from other electrical equipment on the space probe (something similar to the way a magnet can influence another magnet). In this case, the amount of possible disturbances needs to be checked in an electromagnetic compatibility test room.

To make sure that the electrical systems work even in difficult circumstances – for instance, under the influence of the strong electromagnetic radiation near Jupiter – the satellite can furthermore be placed in a powerful electromagnetic field that can cause electrical failures.

Very different tests are required for landers and rovers. To test the parachutes and their deployment, test models of lander probes are sometimes dropped over deserted test areas from aircraft or high-altitude balloons.

Airbags such as those used on NASA's Pathfinder and MER landers need to be tested by dropping them on simulated planetary surfaces. The lower gravity on Mars, 38 percent of that on Earth, can be simulated by letting the probe and its airbags fall on an inclined panel with simulated Martian rocks and dust. The force of the angled impact is then comparable to that of a fall on a horizontal surface on Mars.

Rovers are also tested on simulated surfaces. Engineers have them driven over rocks and through sand and dust with a similar consistency to lunar or Martian soil. These tests can even be useful when the rover is already moving around at its destination: when one of the MER Marsrovers was about to descend inside a crater, a test model was driven through a similar situation in the laboratory on Earth to make sure that its cousin on Mars wouldn't slide down the relatively steep crater wall.

More tests were needed later, when the MER Opportunity became stuck in a soft-sand dune. To investigate the kind of material in which the rover was trapped, and the best way to extricate it, engineers, scientists, outside advisers, and even the project manager at NASA's JPL center mixed sand and powder, dug holes and built dunes. The blend of sandy and powdery materials used needed to match the way the Martian soil had worked itself into spaces between the cleats on Opportunity's wheels. "We tested with one mix of materials, decided it wasn't quite nasty enough, made the mix nastier and tested again. We tested getting the rover stuck and then unstuck in a bunch of different configurations, some of which we think were worse than the one we've gotten ourselves into on Mars," one of the scientists involved explained.

Eventually the team found a way to free the rover, and managed to steer it to more solid ground. The tests were not only crucial for Opportunity's escape, but also for finding out why it got stuck in the first place. This resulted in new rover driving operational rules to prevent similar problems in the future.

Breaking the software

Software testing is a very special activity. A small mistake, such as a missing point or a single wrong number hidden in ten thousand lines of programming code, could be disastrous to an otherwise perfectly designed spacecraft.

Space agencies therefore often commission an independent company – one that has not been involved in the development of the software – to try to find errors. Software validation engineers creatively launch all kinds of possible scenarios to the software to see if it would be able to save the spacecraft out of any situation.

The problem with this kind of testing is that when the verification experts say the software is fine, you cannot be sure that they really tried every possible scenario. If an error is found, you know someone has been looking for it. But how would you know if someone skipped a critical test to be home in time for dinner, and forgot to perform it the next day?

A way to deal with this is to offer an incentive, like a financial bonus if the spacecraft's software operates without major problems for a year or so. This encourages the software testers to really dig deep and find any errors lurking inside the complex programming.

5

INSTRUMENTS OF SCIENCE

IF the mission is not intended to test new technology or merely plant a nation's flag on another planet, the scientific instruments on board a space probe are its most important equipment. They are the "payload" of the spacecraft.

The payload is the reason for the probe's existence; the rest of the spacecraft is only there to take the science instruments to their target, to give them power, to keep them at the right temperature, to aim them correctly and to send their data back to Earth. We use spacecraft instruments in the same way that we use, for example, microscopes in a laboratory on Earth, except that we have to operate them remotely as they are hundreds of millions of kilometers away.

The instruments are what gives us new knowledge about the Solar System, the Universe and thus ultimately about ourselves. And it is a lot. The Voyager 2 mission for instance returned five trillion bits of data from its visits of Jupiter, Saturn, Uranus and Neptune. That's enough to encode over 6,000 sets of the *Encyclopedia Britannica*!

Chapter 5

The More We Learn, the More We Don't Know

All interplanetary spacecraft to date have carried some sort of scientific payload; there is still no other compelling reason to venture into the Solar System except science and curiosity.

The first probes carried simple instruments, as even a fuzzy picture in black and white of the farside (back) of the Moon was quite an achievement in the early days of spaceflight. As science and technology progressed, instruments and space probes became more sophisticated. The more we learned about the Moon or another planet, the more new questions were raised, and this is still so today. Often scientists end up with more questions at the end of a space mission than they had before its launch.

Every interplanetary mission spawns the need for another mission to answer these new questions. Normally it requires more sophisticated instruments than before to find the answers.

Take the Moon, for example. In the early days we just wanted to have a better look at its surface. In particular, the side that is always turned away from us (the farside) begged to be explored, and simple cameras were all it took to do this.

The photos the first Moon probes made of the lunar farside were a surprise. They revealed far fewer of the large, dark, flat areas, called "seas" or "mare regions" than are found on the lunar nearside (the face of the Moon that is turned toward Earth). A new question was raised: Why is there such a difference between the two halves of the Moon?

Moreover, the pictures showed that the lunar surface is covered with craters of all sizes, but could not reveal whether they had a volcanic origin or were created by the impact of meteorites. More detailed pictures by better cameras made during subsequent missions also showed that the whole lunar surface is covered with dust.

How deep was this dust? If we were to send astronauts there, would they sink down in meters of powder? The only way to find out was to land something on the lunar surface. This was much more complicated than just putting a probe in orbit around the Moon, because a lander had to be controlled very precisely to make a soft landing at the right location.

Robotic landers showed that the dust was only a couple of centimeters thick and would cause no danger to a human landing. After a historic effort of human imagination, determination and technology, the Apollo astronauts were sent to the Moon and brought back rocks that showed that the craters were caused by impacts, not by volcanism.

However, even though astronauts walked on its surface, the Moon has still largely remained a mystery. Instruments left by the Apollo astronauts measured seismic activity, which leaves us to wonder whether the Moon is really geologically as dead as we used to think. Also, it is still not really clear what the Moon looks like under the surface.

Robotic probes showed that on the poles of the Moon there are deep craters, the bottom of which may never see any sunlight. In the cold shadow deep down, water-ice brought by impacting comets could have gathered and may still be there. New, sophisticated probes with modern instruments able to detect water-ice will be needed to test this theory.

We have come a long way in exploring the Moon, and we know a great deal more about it than 60 years ago. However, the number of questions we have about it now is a lot higher than the number we started with at the end of the 1950s. Moreover, they are much more complicated to answer.

We discovered that the Moon is a much more interesting place than we thought, begging for further investigation with more, improved space probes and instruments. Not surprisingly, this process of new answers leading to new questions and the constant need for another mission has been consistent for every place we have investigated in the Solar System.

The more we explore, the more we learn how much we don't know about the Universe. The best way to limit the number of unknowns about the Solar System would actually be to stop interplanetary spaceflight at once! However, that would stop the proliferation of questions, but not our curiosity of course. It would also be very boring.

Payload instruments are often rather complex, and of many different types. They are therefore usually developed and built by specialized scientific institutes, laboratories and universities. The scientists responsible for developing an instrument are usually also those who will harvest and analyze the data it sends back.

Only when these experts have crunched the numbers, separated reality from wishful thinking and resolved the mysteries can the discoveries be described in ways that non-experts understand. This process can take months, or even years, from when the original data is collected by the instruments on board the spacecraft. Mission by mission, scientists are building an ever more complete description of the Solar System and the Universe.

Most interplanetary spacecraft have several large and smaller instruments on board. From all the scientific proposals, a good suite of instruments is defined by the responsible space agency; contracts are placed and each instrument gets developed and built by a specific team. In the end all the separate instruments are integrated into the spacecraft.

However, sometimes the payload consists of only one large and very complex instrument, for instance a large space telescope that completely defines the spacecraft layout. The spacecraft may actually need to be built around the instrument, and in that case the instrument and the rest of the spacecraft are often developed in parallel by a large industry, with scientists providing support rather than building the instrument themselves.

In almost any space project there are conflicts between the scientists trying to perfect their instruments, and the spacecraft designers trying to produce the overall most efficient and cost-effective mission design. As already described, the design of a new space mission is always a compromise between conflicting requirements.

LET'S GET PHYSICAL

To understand how instruments on board spacecraft are used, we need to have a look at the physics behind them.

Most spacecraft instruments, especially those on board planetary orbiters, detect electromagnetic radiation in one way or another. Electromagnetic radiation, including visible light, can be described as consisting of waves of electric and magnetic energy (hence the term "electromagnetic"). These waves vibrate up and down at right angles to the direction of motion of the radiation (such as going from the Sun to the Earth) – see Figure 5.1.

When we look for a certain radio station we try to tune the receiver to pick up signals of a certain wavelength. The wavelength for radio signals, and any other type of electromagnetic radiation, is the distance between any two corresponding points on successive waves (for instance, from one peak to the next).

Visible light, the light we can see, ranges in wavelength from about 400 up to 700 billionths of a meter. That's the scale of microbes. There is, however, much more that is invisible to us; the full electromagnetic spectrum ranges from gamma-rays with wavelengths in the range of one billionth of a meter (on the scale of atomic nuclei), to radio waves with wavelengths of centimeters and meters. They are all forms of electromagnetic radiation, which differ from each other in the size of wavelength. (See Figure 1.5 on page 14.)

The frequency of the radiation is directly related to the wavelength, and is the number of waves passing by within one second. It is expressed in Hertz (Hz); 10 Hz means 10 waves per second. As electromagnetic

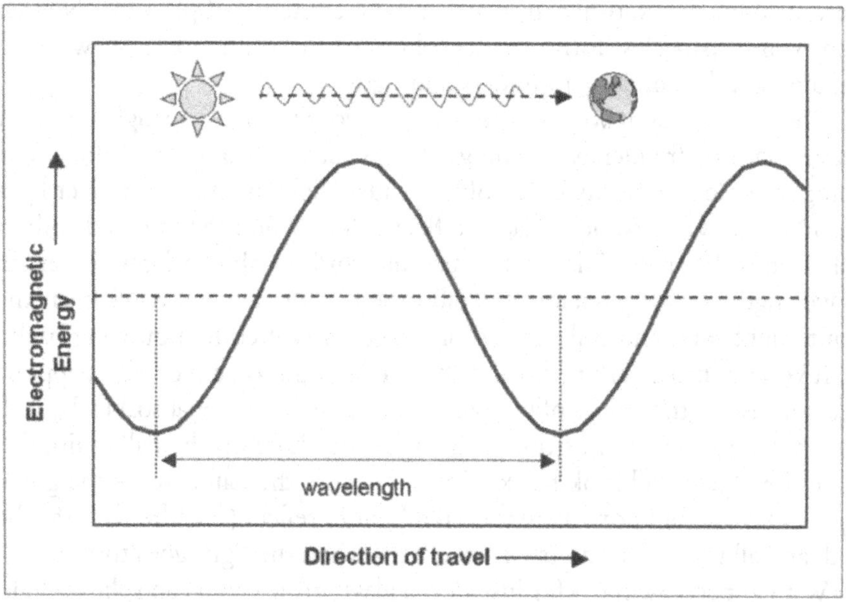

FIGURE 5.1 Light can be described as a wave traveling along the direction of motion.

radiation moves with the speed of light, the longer the wavelength, the longer it takes a wave to pass by, and thus the lower the frequency. And vice versa.

The frequency of visible light, what we call "color," ranges from 430 trillion Hz (red light) to 750 trillion Hz (violet light). Again, the full range of frequencies extends far beyond, from less than 1,000 Hz for very low frequency radio waves, to over 3 billion times a billion Hz for gamma-rays.

Light waves are waves of energy, and the amount of their energy is proportionally related to the frequency (and thus inversely proportional to the wavelength); the higher the frequency, the greater the amount of energy. Gamma-rays thus have the highest frequency and therefore the most energy, which is why they are very dangerous for your health. Radio waves have very low energies; they can be generated by a tiny transmitter with a small battery inside a car-key, for example.

When talking about energy, electromagnetic radiation is often described as consisting of streams of particles or energy packets called "photons" instead of waves. For centuries, the "wave theory" and "particle theory" have been in competition with each other for explaining various phenomena associated with light and other forms of radiation.

Nowadays physicists believe that electromagnetic radiation can behave both as a particle and as a wave, but that either description is only a simplification of something very complex. Some phenomena can best be described by waves, others by assuming the existence of photons. Neither waves nor particles form the absolute truth; both are just ways to mathematically model and understand light.

The "white" light we see from the Sun does not have a single color, or wavelength or frequency or energy, but is made up of many colors. The things we see around us have color because different surfaces reflect light in different ways; some surfaces reflect a lot of blue but no red, others reflect mostly green. This is because the surfaces absorb light of certain wavelengths or frequencies. The absorbed colors are removed from the white light mix, and only the colors left over will come bouncing back.

If you go into a dark room and shine a flashlight on a red ball, it appears red because all the colors of the white light except red get absorbed by it. If you put a green filter in front of the light and shine on the ball again, this time the object will look black. This is because the ball absorbs the green light as before, but now there is no red light to reflect (the filter absorbs the red, and all the other colors except green, when the light goes through it).

With a spectroscope – for instance, a glass prism – you can split sunlight into its constituent colors, like raindrops creating a rainbow. You will then see that sunlight covers the full visible spectrum, from red to violet.

With a good spectroscope, you can also identify peculiar dark and especially bright lines. The dark ones are called absorption lines. The atoms of the Sun's outer atmosphere absorb radiation emitted by the layers below at a series of very specific wavelengths. When this radiation is re-emitted, it is re-radiated in all directions rather than just the original direction of travel. As a result, there are narrow, dark lines in the solar spectrum corresponding to the specific wavelengths of absorption.

Because each type of atom selectively absorbs different wavelengths, we can use the pattern of absorption lines (also called Fraunhofer lines) in the solar spectrum to determine the Sun's chemical composition.

The distinct bright lines in the spectrum are caused by the opposite effect. Here the atoms in the Sun first absorb energy, but then emit this at a series of very specific wavelengths. Instead of blocking light, the atoms are strongly emitting at these wavelengths.

Just as with absorption lines, the emission line patterns form a telltale signature for the type of atoms inside the Sun (the element helium was first discovered in the Sun by spectroscopy before it was found on Earth).

When a molecule absorbs high-energy radiation, and then re-emits it at a lower energy and thus longer wavelength, we call this "fluorescence." So-called Black Lights in discotheques use this phenomenon; they emit ultraviolet radiation that is invisible to us, and gets absorbed by certain materials that subsequently emit the energy as visible light. Although it may seem that white T-shirts and graffiti on walls are freely emitting light on their own, they are merely fluorescing. Again, the wavelengths at which materials fluoresce indicate what they are made of.

Apart from studying the Sun, we can apply all these phenomena to determine what planets consist of. We can study the spectrum of the sunlight reflected by their surfaces and transmitted by their atmospheres; we can see which colors are reflected and which are absorbed by the rocks and the gasses. We can see the wavelengths at which the rocks are emitting themselves. We can also send down beams of radiation energy, such as radar waves, to see how they are reflected or scattered by the surface materials.

IT'S ALL IN THE DATA

The spacecraft instruments that make headlines in the newspaper are nearly always the optical cameras, the imagers that can register radiation at the wavelengths or frequencies of visible light. They give us beautiful pictures of the surface of Mars, or a mysteriously hazy view on what it's like on Saturn's moon Titan. Cameras show us what we would see with our own eyes if we were there ourselves.

However, most interplanetary spacecraft carry many other types of instrument besides cameras for visible light. Often X-ray detectors, radiometers, mass spectrometers and ultraviolet imagers send back more interesting scientific data than the visible light cameras. After all, those imagers only cover the tiny part of the entire electromagnetic spectrum visible to human eyes, while planets radiate at many wavelengths.

Most of the other types of radiation can only be seen by the specialized eyes on our robot probes. Sometimes we can convert that data into visible pictures, similar to a doctor making an X-ray photo of a broken leg. Often such a picture shows something entirely different from how the object appears to us in visible light. A planet or moon may be dark and dull to a normal camera, but glowing in intricate patterns when viewed in infrared or ultraviolet.

However, the data often comes in the form of endless rows of numbers that can only be interpreted by experts with specialized computer equipment. That information may need years of interpretation to be appreciated. However, it may be this stuff that really clarifies how alien some of those other places in the Solar System really are.

Numeric data tells us what the atmospheric pressure is on Mars, and how the surface temperature varies during the day. This gives us a clue to whether we can expect liquid water to run somewhere on the surface, or whether we will only find solid ice underground.

The data may show us how charged particles from the Sun and Jupiter's own moon Io are trapped in the giant planet's electromagnetic field, and how they are bombarding Jupiter's moons. The amount of dangerous radiation determines if we can ever hope to land astronauts there, or whether they would need impractically thick radiation shields to prevent them from being cooked.

Analyzing the material that makes up a comet can help to prove the hypothesis that the building blocks for the complex molecules that make up daisies, dragon flies, dogs, algae and people were brought to Earth by these orbiting chunks of ice. Or it may show that this idea is pure nonsense. One way or another, investigating comets will teach us something about how life on Earth started billions of years ago, and how it may have started on other worlds too. Are comet impacts a vital ingredient or not?

Many Eyes

People and animals have eyes, a nose, ears, and a sense of taste and touch to interpret and survive in the world they live in. Interplanetary spacecraft often have similar senses plus a lot more that are not comparable to anything that exists in the biological world.

Each scientific instrument takes a different look at the same planet, moon, asteroid or comet. Only by combining the data from all the different instruments does the entire picture become clear and we can really get some idea of what we are looking at.

If you receive a bottle of wine without a label your eyes will usually tell you what it is by the size and shape of the bottle. The color indicates whether the wine is white or red. Using your nose you can tell whether it is light and fruity or heavy and oaky. By tasting the wine, you are using the different areas of your tongue that specialize in sensing sweet, sour, salt and bitter to further determine what it is.

By combining all the information, a good wine connoisseur is able to tell exactly the type of wine, where it came from and how it was made. Basing the decision only on taste, or only on smell, or only on what the wine looks like, the expert would seldom be able to make a determination with the same accuracy.

If the expert took the wine into a chemistry lab with its array of specialized equipment, it would be possible to find out even more about the wine, such as the alcohol percentage and whether it contained added substances like sugar or window cleaner.

Planetary scientists are like wine connoisseurs. They need information from all kinds of different instruments to determine the characteristics of an alien world. Therefore interplanetary spacecraft usually carry not just one, but a diversity of imagers and detectors.

Visible light instruments

Visible light instruments include normal cameras that gather light and project it on a light-sensitive surface, just like the camera with which you take your holiday pictures or videos.

In early satellites and space probes, pictures were actually made on conventional film. Earth observation satellites could return that film by launching it back to Earth on board a small re-entry capsule.

As that was much too complicated for interplanetary probes, these robots developed their film on board. This was done by putting a lamp on one side of the film and a light-sensitive sensor on the other side. Dark parts on the negative do not let much light through, while transparent areas do. In this way the film could be scanned by coding the amount of light the sensor received into series of 1's and 0's. The resulting numbers were then transmitted by radio to Earth, where they could be translated back into pictures.

Later, improved systems were based on television camera technology. Images were made up by scanning an object line by line, encoding the data and sending it back to Earth.

Modern space cameras work just like digital cameras that, on Earth, are now becoming more popular than film cameras. The captured light of an image falls directly on a sensor and is immediately digitally encoded. There is no longer a need for film or line-by-line scanning.

The sensor that makes this possible is called a CCD, for Charged Coupled Device. A CCD consists of a vast matrix of detector elements, called pixels. Each pixel converts the energy of the light particles it receives into an electrical charge. The brighter the light that hits a CCD pixel, the higher the resulting electrical charge that accumulates on it.

This electrical charge is subsequently converted into a code of 1's and 0's, so it can be stored in the spacecraft's computer memory. The data is then relayed to Earth by radio, where the signals of all individual pixels are combined according to the matrix layout in the camera. The end result is a complete picture from another world, much like your computer converts coded data in its memory into a picture on its screen.

The individual images are in black and white, because the CCDs only register amounts of light and not its color. However, the cameras can make different images of the same object through different color filters. Using a red filter for instance means that what the CCDs register is the red light from the target; light of other colors cannot pass the filter. Combining the images made through a minimum of three filters (red, blue and yellow) allows us to create a realistic, full color image.

Spacecraft observing planets from space often carry more than one camera. Optical instruments with a wide field of view can see a large part of a planet, but not focus on small details. Cameras with a narrower field of view, called narrow-angle, cover less surface but may be able to show details less than a meter in width.

The broad view, or wide-angle, systems are utilized on the Earth-orbiting meteorological satellites that make the images for the weather forecast on television. On interplanetary spacecraft they are used to capture the results of massive geological forces, such as giant volcanoes, extensive canyons and enormous impact craters.

Those that can focus on small objects – the narrow-angle cameras – can be used to study the edge of a crater, or the blocks of stone that were expelled by the impact of a large meteorite.

When talking about space cameras, the word "resolution" is often used. The resolution is the size of the smallest objects that can still be individually distinguished on a picture made by a camera. For example, a system with a resolution of 2 meters can make out a car from space, but not the smaller object, the driver, standing next to it. With a resolution of a few centimeters you can read a newspaper's headlines from orbit (apparently modern Earth-orbiting spy satellites are able to do this, but no one working with them is allowed to admit that, of course).

In CCD cameras, the resolution is identical to the size of an object represented by one pixel on the digital photo. When you zoom in on such a picture, you'll see that it is built up of individual blocks. Each block represents a pixel and thus objects smaller than such a block cannot be seen as individual items.

Another important parameter for a camera is its sensitivity – a measure for the weakest light source it is able to detect. It's similar to the ISO number for the film in conventional cameras. The higher the sensitivity, the less light is needed to make an image.

The nice thing about visible light cameras is that they give us a look at a planet as if we were actually there. You don't need to be a trained scientist to recognize a huge canyon on a picture of Mars or a crater on the Moon. Because of this, the pictures in visible light are the ones that end up in newspapers and on the covers of popular magazines. They are turned into posters, downloaded to serve as wallpaper background for computers screens and decorate mouse pads, T-shirts and coffee mugs.

With two cameras looking at the same thing under slightly different angles we can make stereoscopic images. When looking through a viewer that shows one image to your left eye and the other to your right eye, you get a perception of depth in the image. This is a great way to get a better understanding of a planet's landscape in three dimensions. Moreover, by using both images you can calculate the height of mountains and other features in the picture.

Stereoscopic images make for great posters and popular books as well; the three-dimensional aspect can be cheaply reproduced by printing each picture in different colors and supplying a pair of glasses with one green and one red lens. The green lens allows only green light to pass, so that one eye sees only the green image and, similarly, your other eye sees only the red image. In this way each eye sees a different picture with a slightly different view angle, resulting in a three-dimensional image.

NASA's Sojourner Marsrover and its two recent Mars Exploration Rovers had stereo-camera systems mounted to obtain 3-D images of the Martian landscape.

ESA's Mars Express orbiter also makes stereoscopic images, but uses only one camera. As it flies over an area, it looks at features on the ground from different angles at different times. Combining the views with a

FIGURE 5.2 The High Resolution Stereo Camera of Mars Express. [EADS Space]

computer on Earth, it is possible to create very rich stereoscopic images with an unprecedented amount of detail. Mars Express is the first spacecraft to apply this novel technology.

Apart from making great pictures, visible light can also be used to make spectroscopic measurements. With a spectroscope you can see the overall spectrum of radiation an object reflects or emits, and the absorption and emission lines that indicate which atoms are present. With this information, you can thus find out the material something is made of, purely by the light you receive from it.

For instance, we know very well what sunlight reflected by water looks like because we can study that here on Earth. If a satellite in orbit around another planet detected a spectrogram identical to that of water on Earth, we would know we have had found water on another world.

There is more than meets the eye

Cameras for visible light are great scientific instruments but they cannot see everything. First, they depend on reflected sunlight. They cannot see through clouds and murky atmospheres, and at night they cannot see anything. Rovers or landers on the surface may take their own lights with them, but you cannot light a whole planet from space. Second, with visible light you cannot see what is under the ground, or whether the ice on the Martian polar caps is made of water or frozen carbon dioxide.

Objects with a temperature above absolute zero, and that is everything in the known Universe, emit radiation over a wide range of wavelengths. However, the wavelength at which an object emits the bulk of its energy depends on its temperature. Relatively cold objects radiate a lot in the infrared, while objects at extremely high temperatures mostly emit X-rays and gamma-rays. Visible light is thus not always the best way to look at something; an object that appears dull to our eyes may shine brightly at wavelengths that are invisible to us.

Infrared systems, for instance, detect radiation with longer wavelengths than visible light. In visible light we see objects by the sunlight they reflect, but in infrared we can see them by their self-emission, i.e. the radiation the objects give off because of their temperature. People and animals keep themselves warm and therefore usually have a higher temperature than their surroundings. Because of this they emit more infrared radiation and thus show up brightly on infrared camera images. Infrared detectors thus also work at night, which is why on Earth we use them in security systems for houses, shops and offices.

For these reasons spacecraft observing planets, comets and asteroids typically carry a range of instruments with them, each specialized in detecting radiation within a certain range of wavelengths or at a particular wavelength.

In infrared you can, for instance, measure how a planet's surface cools off at night. This is very interesting, because it can say something about its composition, or how well its atmosphere acts as a blanket.

Microwave instruments detect radiation with wavelengths between those of the far infrared and the shortest wavelength radio waves. You cannot individually distinguish objects that are smaller than the wavelength of the radiation you are measuring, so the resolution of microwave instruments is less than for visible light imagers. However, they can collect unique information over very large areas.

Passive microwave radiometers detect radio waves from the Sun that are reflected by a planet. They are mainly used on Earth-observation satellites, for measuring the sea surface temperature and sea ice coverage, and to get information about the amount of water on the surface and in the ground. Microwave radiometers can even be used to measure wind velocities, as it was discovered that wind changes the way radio waves are reflected by the sea.

Radar systems are active forms of instruments, because they generate the needed radio waves themselves and then look at how a surface or object reflects that radiation. Radars thus do not need any sunlight or radiation emitted by an object itself. The principle is the same as in radar systems on board ships and airplanes.

Radars are very powerful instruments. For example, they are capable of looking through cloudy atmospheres and some can even look through solid material – similar to the way radio waves can go through the walls in your house so that the radio also works indoors. With certain types of radar you can see what is underground, beneath the planet's surface, because each layer in the ground will reflect some of the radio waves.

With radar you can also measure the distance between your spacecraft and the surface below. Because the radio waves travel with the speed of light, you can calculate the distance by determining how long it takes the waves to reach the surface and reflect back to the instrument. Radar can thus be used to measure the relief or "topography" of a planet's surface: the height of the mountains and the depth of the craters.

Moreover, the roughness of the surface of a planet influences the way in which the radar waves are reflected back to the receiver. A smooth ice plane reflects the waves back without much disturbance, but a rough and rocky surface scatters a big part of the waves in different directions. With a radar you can thus not only gauge the topography of a planetary surface, but also the overall properties of the surface materials.

NASA's Magellan spacecraft used a sophisticated radar to make detailed maps of the surface of Venus, which is hidden from us by a very thick, almost opaque atmosphere ("almost," because visual and ultraviolet radiation are blocked, but some infrared radiation gets through). The instrument combined the unique features of radar by peering through the murky atmosphere and also measuring the topography of the terrain. From 1990 until 1994 it orbited Earth's sister planet and gave us the most detailed map of the surface of Venus ever.

The MARSIS experiment on board ESA's Mars Express involves three long antennas that send low-frequency radio waves toward the planet it orbits. Most of the radio waves bounce back from the surface of Mars, but a significant fraction travels through the crust. This energy is reflected where layers of different materials meet – for instance, at the border between water and ice.

MARSIS maps the planet's subsurface structure to a depth of a few kilometers. It is able to measure the thickness of the sand in dune areas and where layers of sedimentation are covering other material. MARSIS also studies the ionosphere of Mars, an electrically charged region of the upper atmosphere that also reflects some radio waves.

After the launch of Mars Express, new computer analyses indicated that during the deployment process the long antennas could possibly whip against the spacecraft and might damage it. The deployment of MARSIS was therefore delayed until sufficient data from the other instruments had been collected and the problem had been studied more carefully.

Nearly one and a half years into the mission, it was decided that the time had come to extend the antennas. However, after the deployment of the first 20-meter (66-foot) antenna, analysis by flight controllers at ESA's European Space Operations Center in Germany showed that one of the outermost segments had deployed but was not locked into position.

Engineers speculated that the long storage in the cold expanses of space could have affected the fiberglass and Kevlar material of the antenna boom. The mission team therefore decided to swing Mars Express so that the Sun could heat the cold side of the antenna. They hoped that as the cold side expanded in the heat, it would force the unlocked segment into place.

After an hour, Mars Express was pointed back at Earth so that its high-gain antenna could be used for communications again. The data it sent showed that all antenna segments had now successfully locked.

Just over a month later, and after careful analyses, the second radar boom was deployed without a hitch. This time, ESA controllers had put

Mars Express into a slow spin to ensure that the boom and its hinges would be evenly warmed by the Sun as it extended. The shorter third and final antenna was also deployed a few days later, and MARSIS could finally start its work.

Its first findings included buried impact craters, layered deposits at the north pole and hints of the presence of deep underground water-ice. Layers of lava that cover some of the buried crater walls indicate that Mars was once a geologically active planet.

A new way of measuring the distance between a spacecraft and the ground is by use of laser altimeter instruments. These fire short pulses of laser light down to the surface and record the time it takes to receive a reflected signal. They are more accurate than radars because they operate at shorter wavelengths and have more focused beams, but cannot see what is beneath the surface.

There is a range of other instruments an orbiter can use to study a planet. As an example, let's first have a look at what is on board Messenger, the NASA spacecraft that is on its way to Mercury now.

MESSENGER'S INSTRUMENTS

Messenger's suite of diverse scientific instruments is designed to find out as much as possible about Mercury, the planet closest to the Sun.

We already know that Mercury has an unusually high density, indicating that it has a large metallic core (heavy metals tend to sink down) accounting for about two-thirds of the planet's mass, covered with a layer of rocky material (silicate) with a lower density. The metallic core probably has a diameter about three-quarters of that of the planet, which makes it proportionally the largest of any known planetary body. The Earth's metallic core is much smaller in relation to the size of our planet, comprising only about one-third of its total mass.

There are some theories that try to explain why Mercury has such a large core, or, from an alternative perspective, such a thin silicate mantle. One suggests that Mercury was initially created more normally, but a collision with another planet blasted off most of its outer layer. Or maybe the young, extremely hot Sun evaporated a large part of the mantle. On the other hand, Mercury may have simply been born the way it is. Messenger's instruments are designed to help us to find out what really happened.

The spacecraft carries spectrometer instruments that are able to measure how different areas on Mercury emit X-rays, gamma-rays and neutron particles.

Elements such as oxygen, silicon and iron emit gamma-rays when hit by cosmic radiation coming from deep space. At the same time, certain naturally radioactive elements emit gamma-rays by themselves. All this will be measured by Messenger's Gamma-Ray and Neutron Spectrometer, or GRNS, instrument. It will be able to measure what the surface consists of to a depth of about 10 centimeters (4 inches).

GRNS will also detect low-energy neutron particles that are created when cosmic radiation hits hydrogen-rich material, down to about 40 centimeters (1.3 feet) below the surface. This makes the instrument useful for finding water-ice at the poles of Mercury, were it may be hiding in the cold shadows inside craters where sunshine never manages to reach.

The X-Ray Spectrometer, or XRS, measures emissions at X-ray wavelengths that are caused by radiation from the Sun that strikes the planet's surface. X-rays from the Sun cause atoms in the Mercury surface to fluoresce and emit X-rays of their own. The precise energy carried by these X-rays is a signature of the element emitting it. XRS will give similar information as the GRNS instrument, but only detects the elements in the upper 1 millimeter (0.04 inch) of the surface.

Together these spectrometers will thus be able to tell us what elements can be found on Mercury, how they are spread over the planet's surface, and how their abundances differ between the visible first upper millimeter and the layers lower down.

The Mercury Atmospheric and Surface Composition Spectrometer, MASCS, is an instrument that is sensitive to light from infrared to ultraviolet wavelengths. Apart from also helping to find out the kind of minerals that are present on the surface, it will be able to detect the various gases the extremely tenuous atmosphere of Mercury contains.

Employing all these instruments, Messenger will provide scientists with a map of the elements that are present on the surface, just under it, and in the atmosphere. This geological and chemical information is vital for checking the credibility of the various theories that exist about the formation of Mercury billions of years ago.

Since about 55 percent of Mercury's surface has never been photographed at close range, one of the main mission objectives of Messenger is to map the entire planet. We will see areas that the Mariner 10 spacecraft was not able to shoot during its short flybys in the mid-1970s. Moreover, we will see these and the already known part of the surface with much more detail than ever before.

The Mercury Dual Imaging System, MDIS, consists of a wide-angle and a narrow-angle imager. The wide-angle camera also incorporates a wheel with filters that can be rotated into its field of view one at a time. Using these, the instrument can be used for spectroscopic imaging, because each filter only lets through light with a particular wavelength. The wide-angle camera can therefore give additional information about what the surface of Mercury consists of.

MDIS will also be used to make stereoscopic images of Mercury by combining photos made of the same area from different positions in space. This will show the relief of the surface in better detail than normal pictures, because it will provide a 3-D view of the planet.

To map the relief on Mercury, Messenger furthermore incorporates an altitude-measuring instrument called the Mercury Laser Altimeter, MLA. This contains a laser that sends pulses of laser light down to the planet's surface and a sensor that collects the light after it has been reflected. The instrument measures the amount of time it takes the light to make a round-trip to the surface and back, enabling scientist to determine the height of features on the surface. With this information, and accurate knowledge of the spacecraft's orbit, they can then make very exact maps of Mercury's topography.

Our Earth has a strong magnetic field because of the movements in its liquid metal core. Mercury's core is much smaller in actual size (but not as a percentage of the total mass) and therefore should have cooled down and become solid a long time ago. A solid metal core does not create a magnetic field, but the Mariner 10 spacecraft nevertheless found one at Mercury. The core may somehow have remained liquid, or the current field may be the original field's remains, still left "frozen" in the now solid core.

A magnetometer has been put on board Messenger to measure the strength and direction of the magnetic field. This instrument will map Mercury's magnetic field and will search for regions of magnetized rocks in the crust. To minimize the disturbance by the spacecraft's own magnetic field caused by its electrical components, the magnetometer is placed at a distance, at the end of a 3.6-meter (12-foot) boom.

Since Mercury has a magnetic field, it also has a magnetosphere. A magnetosphere is the region surrounding the planet in which electrically charged particles are controlled by the magnetic field of the planet rather than by the Sun's magnetic field, which is carried by the solar wind. Planetary magnetospheres can attract and trap charged particles coming from the Sun and interstellar space.

The Energetic Particle and Plasma Spectrometer, EPPS, will measure the composition, distribution and energy of these charged particles (electrons and various ions) in Mercury's magnetosphere.

The Radio Science experiment does not use a specialized instrument, but makes use of Messenger's normal radio communication system. When the spacecraft moves away from us, the radio waves we receive will be "stretched" and we will therefore detect the signal on a slightly lower radiofrequency. When Messenger moves toward us, the radio waves will be "compressed," resulting in a slightly higher frequency. This is called the Doppler effect.

By measuring this effect, we can detect slight changes in the spacecraft's velocity as it orbits Mercury. These changes are the result of the fact that the gravitational field of Mercury is not completely homogeneous. Some parts of the planet have a bit more mass and thus a slightly greater local gravitational pull than other parts. As a result, spacecraft orbiting Mercury will not move in perfectly smooth orbits, but will be disturbed in their movement. Studying the Doppler effect on Messenger's radio signals allows scientists to learn about Mercury's mass distribution, including the variations in the thickness of its crust.

Roving Around

Rovers driving around on a planet's surface need to have different instruments to orbiting spacecraft. Rather than seeing large areas from high above, they investigate soil and rocks on the spot.

Rovers resemble animals more than orbiting spacecraft, and often have a long boom "neck" that enables the cameras to be high enough to get a good overview of the terrain. The cameras have multiple "eyes" to give them a stereoscopic view and a sense of depth. Some rovers also have arms that can reach out to inspect and sample rocks and soil.

Because of their mobility and likeness to living beings, rovers are popular topics for toys. In 1997, after the landing of the real one, the Mars company gave away small Sojourner Marsrovers with packs of their candy bars. More recently, the LEGO company issued a Mars Exploration Rover with functional deployment systems, wheels, steering mechanism, moving antenna and camera head, and even an extendable experiment arm.

NASA's Mars Exploration Rovers (MERs) Spirit and Opportunity, the inspiration for the LEGO toy, landed early 2004 on different sides of the planet. Their sets of instruments were especially selected to determine whether Mars was once a more friendly place where life may have evolved. The evidence was expected to be found in the rocks, of which the

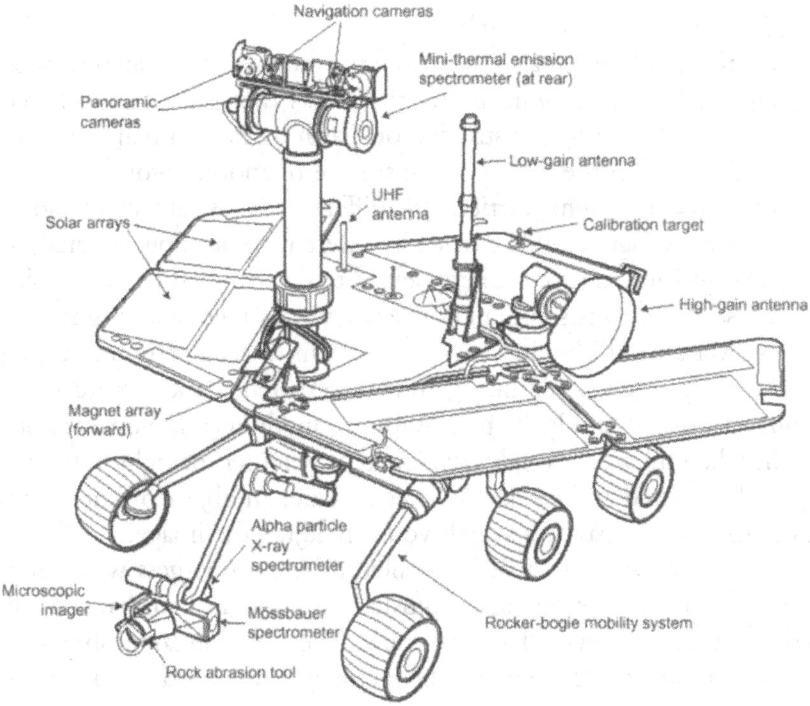

FIGURE 5.3 NASA's Mars Excursion Rovers have arms to investigate rocks and a stereo-camera system on a long neck. [NASA]

composition would hopefully betray whether they had formed in a watery environment a couple of billion years ago.

Each identical rover contained a panoramic camera and a spectrometer on the rover's body, and a set of instruments on a mechanical arm to analyze rocks and soil. A grinder on this arm could be used to scrape away the outer layers of rocks so that the pristine material inside could be investigated.

The panoramic camera of each rover consisted of two high-resolution stereo cameras on top of a mast in front of the vehicle. Positioned 1.5 meters (5 feet) above ground level, they acted as the eyes of the robot and made it possible to create three-dimensional panoramic images of the terrain.

The images the two golf-cart-sized rovers sent back to Earth showed an unprecedented amount of detail. We saw vast, wind-swept landscapes of red and brown rocks covered in rust-colored dust, large craters and exposed layers of sedimentation that were probably deposited by liquid water. They showed sunrises and sunsets through the orange atmosphere, and dunes of fine dust. Looking through the eyes of the rovers, you can imagine being on Mars yourself.

Parts of the MERs are also shown on the pictures. You can see the solar cells and the bolts, the wheels and the tracks they left behind. It seems fantastic that something constructed on Earth by human hands is now so far away, placed on the red, desolate surface of another world!

Each camera system carried 14 different types of filters, so that spectroscopic images could be made in different colors. Spectral analysis of the rocks, soil and the atmosphere helped to determine their composition.

Because the scientists didn't know exactly what color and brightness the light on Mars would be, it was hard for them to accurately fine-tune the pictures they receive. It's like getting a very red, dark picture of your friends made in a dimly lit bar. Without anything else you cannot say whether the film was wrongly developed or whether the light in the room was really that red and weak. You need a clue, an object of known color on the picture – a base on which you can adjust the image.

Therefore the rovers carried so-called calibration targets with patterns of known colors. The scientists adjusted their pictures so that the colors on the pictures matched what they knew the calibration target should look like. As a result, all the colors on the pictures of the Martian surface could also be corrected to be realistic.

When the first pictures from the Viking landers came back in the 1970s, photo interpreters incorrectly assumed that the sky on Mars would look as it does on Earth. Therefore they adjusted the colors on the first pictures in such a way that the sky looked blue. However, once the pictures were correctly calibrated, they realized that the Martian sky was actually a shade of pink, because of all the rusty dust it contains!

The MERs carried panoramic camera calibration targets in the shape of a sundial. The corners of the sundial had colored blocks for adjusting the camera's sense of color. The shadows cast by its center post allowed scientists to correctly scale the brightness of each image.

Moreover, to get young people involved in space exploration, children were encouraged to track the time on Mars using the sundial as part of NASA's Red Rover Goes to Mars education program.

Another instrument on the rover's bodies was the Mini-Thermal Emission Spectrometer, which could see the infrared heat signature emitted by Martian surface features. It could determine the mineral composition of soil and rocks from far away. The spectrometer could even see through the dust that coats most of the rocks on Mars.

The main part of this instrument was housed inside the rover, but the mast for the cameras was used as a periscope so that it could get a good view of the terrain: a mirror assembly on top of the boom reflected the infrared light from outside down through the mast into the spectrometer.

With this instrument, scientists and operators could determine whether it was worth taking the rover in for a closer look, or better to move on and find a geologically more interesting location. The spectrometers on the two rovers were also pointed upward to make the first-ever high-resolution temperature profiles through the atmosphere of Mars.

For detailed measurements, the rovers had an instrumented mechanical arm on the front that could be extended against rocks and the ground. It also incorporated a Rock Abrasion Tool that could grind dust and weathered outer layers of rocks, so that the pristine material inside was exposed for investigation. It was the equivalent of a geologist's hammer, which is used to break rocks to be able to see what they're made of.

The Microscopic Imager on the arm, a combination of a microscope and a camera, was used to make extreme close-up images in black-and-white of the rocks and soils. The imager could see tiny grains of minerals embedded in the rocks, and measure the shape and size of particles in the Martian soil. It was hoped that it would also be able to see fossils of microbes, but none was found.

Iron is a very common material on Mars. In fact, the surface looks red because of the oxidized "rusted" dust. The scientists thus needed an instrument specifically designed to determine the composition and amount of iron-bearing minerals and the magnetic properties of the iron. With this information, they learned a lot about what Mars was like when these minerals were formed.

The instrument that could do this was a Mössbauer Spectrometer. Germany provided two of these to NASA for its Mars Exploration Rovers. The spectrometer used two tiny pieces of radioactive cobalt-57 to radiate the samples with gamma-rays. The spectrometer then analyzed how the radiation got absorbed by the sample to gain information about the iron (the absorption is known as the "Mössbauer effect," hence the instrument's name).

The Mössbauer Spectrometer on the rover Spirit found an iron-bearing mineral called goethite in a rock in the Columbia Hills of Mars. As this mineral is normally produced only in the presence of water, the examination of the rock produced strong evidence that there was once water in the area that Spirit explored.

The scientists called this particularly interesting rock "Clovis." It is a geologists tradition to give interesting sites and rocks a name, to enable them to be easily referred to. Names are easier to remember than boring numbers and codes.

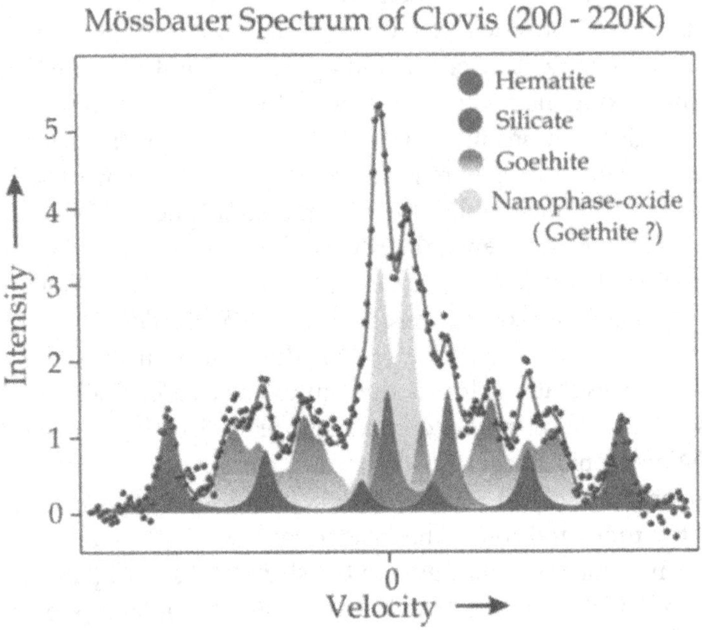

FIGURE 5.4 Mössbauer Spectrogram of the "Clovis" rock, which turned out to contain a lot of water-formed Goethite. [NASA]

The rover arm also included another type of spectrometer, called the Alpha Particle X-Ray Spectrometer, which was also built in Germany. It works somewhat similar to a Mössbauer Spectrometer, but uses Alpha particles (radiation consisting of subatomic fragments with two protons and two neutrons) and X-rays emitted by small amounts of radioactive curium-244. It was used to determine which other elements make up the rocks and soil on Mars.

Each rover also carried a set of three Magnet Arrays, supplied by Denmark, to collect dust. There is a lot of dust on Mars and, because of the high iron content, some of it is magnetic. The magnetic minerals in the fine dust may be freeze-dried remnants from the time that Mars was a much wetter planet.

One set of magnets was put on the Rock Abrasive Tool, so that it could gather the dust that resulted from the tool grinding through the outer layers of the rocks. By looking at the amounts and the patterns the dust made as it accumulated on magnets with different strengths, scientists got indications of the composition of the dust, which includes iron.

Another set of magnets was mounted on the front of the rover to collect airborne dust. The dust collected on these magnets could be reached by the mechanical arm's Mössbauer and Alpha Particle X-Ray Spectrometers, which could also be used to analyze the Martian dust.

A third magnet was mounted on top of the rover and was powerful enough to collect dust blown over the vehicle by the wind. The dust on this magnet could be viewed by the panoramic camera.

In addition to the main suite of instruments, some equipment that was mainly used for engineering and operational reasons could also be employed to give a better idea of the geology of the area.

The Hazard Identification Cameras, mounted low on the front and back of each rover, provided 120-degree stereoscopic black-and-white images. These were used to help to find obstacles in the vicinity of the rover, and for steering the rover's mechanical arm with its abrasion tool and instruments.

The Navigation Camera also consisted of a stereo pair of black-and-white imagers. It sat on top of the mast, together with the Panoramic Camera. However, it could make pictures with a much wider angle, to give quick full-circle views of the rover's location. If the Panoramic Camera was used for this purpose, the rover would need to send a lot more data to Earth and would therefore take much more time.

The rover's wheels were, of course, mainly used for mobility, but could also dig shallow trenches to show soil properties and expose fresh soil that could then be investigated by the instruments.

The two rovers each landed in very different terrain. Spirit came down in the Gusev Crater, and found it to consist of volcanic rocks that had obviously been little affected by water. However, when it started to clamber into the Columbia Hills, it found rocks that had been ground into small grains, then glued back together with salt. This salt must first have been dissolved in liquid water to get between the grains, and when the water evaporated the salt cemented the grains to each another.

Spirit also found the earlier mentioned concentrations of goethite mineral. Nevertheless, the rover provided no evidence that large amounts of water had ever been present for extended periods.

In contrast, on the other side of the planet, Opportunity saw terrain that had clearly been shaped by lots and lots of water. Near the small crater in which it had landed, it spotted an outcropping of layered rocks that was clearly built up of sediments deposited by water. Moreover, these rocks were found to contain very high amounts of sulfate, indicating that they had been formed by the evaporation of sulfur-rich, standing water.

Opportunity's cameras then made close-ups of curious, small stony balls embedded in the rock and scattered all over the ground. The geologists

FIGURE 5.5 "Blueberries" cover the ground in this image made by Opportunity. The circular area on the rock has been cleaned with the Rock Abrasion Tool for inspection with the arm's instruments. [NASA]

quickly dubbed them "blueberries." The instruments on the rover's arm determined these to consist of hematite. While hematite can be deposited by volcanic lavas, it is often formed in the presence of water.

The MER discoveries, together with the global data obtained from the various Mars orbiters, seems to indicate that Mars was once covered with rivers, lakes and perhaps oceans. However, water may have only occasionally and briefly flowed through them. Mars seems to have been mostly a frozen planet for the last couple of billion years. Although it may never have looked like Earth, the planet probably once looked a lot more friendly than it does now.

Operating all the scientific equipment on board the rovers was quite a job. During their main mission of three months, the MERs required a science team of 50 people and an engineering team of 30, equally distributed over the two rovers. The 40 team members per rover worked in two shifts.

At the beginning of each Mars day the scientists would meet to decide on the new activities, according to the mission plan, past experience and the situation in which the rovers were left the previous evening (during the Martian night, without sunlight for the solar arrays, the MERs were "sleeping"). Their decisions were then conveyed to the engineering team, who were tasked with preparing the commands and relaying them to the rovers.

The operators often had to work at very strange times, because their shifts were not following normal time on Earth, but were dictated by the local time on Mars. When it was midday at one of the rover's location, it could be in the middle of the night at the NASA operations center. Moreover, the two rovers were driving around on opposite sides of Mars, so while it was daytime for the Spirit rover team, it was night-time for those working with Opportunity.

In addition, a Martian day, called a "Sol," is equivalent to roughly 24 hours 37 minutes on Earth. As a result, the team's working hours were moving 37 minutes later each day with respect to the time on their Earth clocks.

Initially operating the delicate and expensive rovers was done one step at a time. A typical sequence would go something like this. On Day 1, the rover is commanded to drive a meter or so toward a preselected rock and take a photo for navigation purposes. The view on the picture and the overall situation are then evaluated in order to prepare the commands for Day 2, during which the MER is told to drive a bit closer and take more pictures. On Day 3 the Rock Abrasive Tool is put against the rock to scrape away the outer surface, and a picture is taken of the resulting cleaned spot. On Day 4 the Microscopic Imager and the spectrometers on the arm are employed to study the exposed, fresh material. On Day 5 the results are studied and if the data is satisfactory the MER can be commanded to move on, but if not, then additional measurements may need to be taken. On Day 6, the rover might then be instructed to prepare to head for a new target.

The data we obtained from the MER instruments tells us that their landing sites, especially those of Opportunity, are full of clues to a watery past. The rovers found minerals that only form in the presence of water, and saw rocks that showed fine layers that were truncated, discordant and at angles to each other typical of sediments laid down in flowing water. It now seems certain that large amounts of liquid water must once have existed on the surface of Mars.

The big questions remaining are how long Mars existed as a relatively wet (and warmer) planet, and whether its existence was long enough for life to have evolved. If so, did life also actually *start* there or not? If it didn't, we should ask ourselves why it happened on Earth and not on Mars. The riddles of the red desert planet are bound to keep us busy for quite some time!

DIVING THROUGH AN ATMOSPHERE

A breed apart are probes that are not designed to orbit a planet or to land on it, but primarily to investigate its atmosphere. Descending under parachutes, these probes soak up atmospheric gasses for analyses, and measure temperatures, pressures and densities as they go down.

Naturally these missions don't last long, so all payload instruments have to work flawlessly from the moment the protecting heatshields fall off. There is no margin for error, and no time to fix bugs or do experiments another time. It's a "one chance only" situation.

In 1995 NASA's Galileo orbiter arrived at Jupiter. It carried a 339-kilogram (747-pound) probe that was ejected to descend into Jupiter's atmosphere. Once decelerated to a mere 100 kilometers (62 miles) per hour by the atmospheric drag, the probe ejected its protecting heatshields and continued to descend for about an hour, measuring gas compositions, temperatures, pressures and wind speeds. At a depth of 130 kilometers (80 miles) the short but exciting mission ended abruptly when the Galileo entry probe was destroyed by a pressure of 25 times that on the Earth's surface and a temperature over 150 degrees Celsius (300 degrees Fahrenheit).

NASA and ESA are studying probes that may venture even deeper into Jupiter's atmosphere than Galileo. However, no probe will ever be able to reach the heart of Jupiter, as the pressures and temperatures there are much too high for any type of machine.

In 1997 NASA, ESA and the Italian Space Agency launched their combined Cassini–Huygens mission, the most ambitious effort in robotic planetary space exploration ever. Its target was the planet Saturn, the "Lord of the Rings." It would take no less than seven years and a series of flybys of Venus, Earth and Jupiter to arrive at its destination.

The giant gas planet Saturn, second in size only to Jupiter, consists mostly of hydrogen and helium and shows intricate bands of clouds in various beautiful shades of brown and yellow. However, it is most famous for its complex system of bright rings made of rock and ice particles, ranging in size from grains of sand to as large as a truck.

The NASA orbiter Cassini, named after the famous Italian-French astronomer who studied Saturn in the seventeenth century, was designed to get into orbit around Saturn and investigate the planet and its beautiful rings from close up. The continuously changing orbital inclination of Cassini allows it to explore the environment around the planet over a wide range of latitudes. Moreover, its varying orbit allows the spacecraft to roam around the planet's fantastic collection of strange and fascinating moons.

FIGURE 5.6 Artist impression of the Cassini spacecraft releasing the Huygens probe. [ESA]

Cassini was also acting as a mothership for the ESA probe Huygens, named after the Dutch astronomer Christiaan Huygens, who discovered Titan in 1655. Huygens was designed to investigate the thick atmosphere of Saturn's largest moon, Titan.

The probe was decoupled from Cassini nearly six months after arrival at Saturn, because only then would the orbits of the orbiter, the ejected probe and Titan all be aligned in the right way. Then the 320-kilogram (700-pound) probe entered Titan's atmosphere while protected by a large heatshield made of carbon fiber. Once the velocity had diminished sufficiently, parachutes further slowed the probe down and let it drift toward the surface.

While measuring important properties of the atmosphere, Huygens sent its findings to Cassini, which stored the data on its onboard computer and then forwarded it to Earth.

Although not specifically designed as a lander, Huygens was expected to reach Titan's surface and give us the first views ever of what it looks like. And it did.

Primeval soup

No moon in the Solar System is more mysterious than Titan, which is actually larger than the planet Mercury and our own Moon. It is the only moon with a substantial atmosphere, consisting mostly of nitrogen but laced with a mixture of gases called hydrocarbons, especially methane.

Titan's surface cannot be seen from space; it is hidden behind a brownish-orange haze. This mist consists of complex organic molecules (molecules based on chains of carbon atoms, like those in our body) that are created in the upper atmosphere. Under the influence of the Sun's ultraviolet radiation, methane molecules are broken apart and recombine in the form of new organic molecules. The smoggy, globe-enshrouding haze forms clouds that resemble car exhaust fumes, from which scientists expected a rain of complex molecules to be falling down to the surface.

Far from the Sun, Titan is thought to be a deep-frozen memory of the early Solar System. Scientist think that the conditions on its surface and in its atmosphere make it much like an early Earth of around 4 billion years ago, but in deep freeze, before life began to evolve and filled the atmosphere with oxygen. Titan's brownish veil of chemicals and the organic goo raining down on the surface may resemble the primeval soup in which the first living organisms on our planet once appeared.

Earlier, information on Titan had been collected by powerful telescopes on Earth, space telescopes in Earth orbit and most importantly the Voyagers 1 and 2 spacecraft that, respectively, flew by in 1980 and 1981. The data indicated that the pressure at Titan's surface must be about 1.5 times that on the surface on Earth, and that the temperature there is about –180 degrees Celsius (–290 degrees Fahrenheit).

However, apart for some idea of the composition and average surface pressure, not much was known about Titan's atmosphere. Were the scientific models that predicted liquids of complex organic molecules on the surface correct? Could it really rain on Titan? To answer these questions, and others, scientists equipped Huygens with a range of instruments to gain as much data as possible during its short descent.

Going down

Let's go back to December 2004 and follow Huygens as if we are flying along with it, starting just before its epic descent into the mysterious atmosphere of Titan.

On Christmas Day 2004, after a final system check, Huygens is ejected from the Cassini orbiter. To spare the batteries, everything on board the probe is turned off, except for a timer system to activate the sleeping robot when the moment of truth comes. For 22 days Huygens journeys quietly and passively toward mystifying Titan.

Then, on January 14, it gets its wake-up call from the timer system. Huygens comes alive, feeding power from the batteries to all its subsystems and scientific instruments. Hours later the giant Green Bank radio telescope in the USA is the first to pick up the faint beacon signal of Huygens directly. The signal is far too weak to carry any data, but it is the earliest indication that Huygens works. Hearing Huygens directly is quite a feat, similar to receiving a mobile phone signal from Saturn.

Four-and-a-half minutes after activation Huygens slams into Titan's upper atmosphere at 65,000 kilometers (40,000 miles) per hour, fast enough to fly across the USA coast to coast in about 2.5 minutes. Within 3 minutes the atmospheric drag reduces the speed to 1,500 kilometers (930 miles) per hour. The huge amount of energy that the capsule has in terms of velocity (kinetic energy) is mostly transformed into heat. It makes the temperature of the front of Huygens' protective shield race up to 11,000 degrees Celsius (21,000 degrees Fahrenheit), about twice the surface temperature of the Sun.

Then, after having been packed inside its container for over nine years, a small pilot chute comes out and pulls away the aft protective shield that is covering the probe itself. Almost immediately, still 180 kilometers (110 miles) above the surface, a large parachute unfurls to further slow down the falling explorer. On Earth the atmosphere only starts to be noticeable at an altitude of about 100 kilometers (62 miles), but Titan's atmosphere is much thicker.

Thirty seconds after opening, the main parachute has brought the remaining velocity down to 320 kilometers (200 miles) per hour, and the front protective shield is ejected. The parachute is relatively large to ensure that Huygens descends much more slowly than the falling shield. This prevents the probe from accidentally slamming into it.

Now Huygens is completely free, and its instruments start their investigation of the atmosphere. The data is sent to Cassini, which is passing by Titan at a distance of 60,000 kilometers (37,000 miles). Cassini will store all the information, then dispatch it to Earth via a much more powerful radio signal than the Huygens probe can provide.

For 15 minutes Huygens gently glides down under the large parachute. The atmosphere is getting thicker and thicker, further slowing down the probe. Soon the rate of descend is so low that if Huygens continues like that the batteries might be empty before it reaches the surface. Therefore the main parachute is released at an altitude of 138 kilometers (85 miles) and a smaller drogue chute with less drag takes over. A bit faster, Huygens continues the descent through the Titan atmosphere for the next two hours, its instruments busy with collecting valuable scientific information.

NASA's Gas Chromatograph and Mass Spectrometer (GCMS) instrument identifies and quantifies the gasses comprising Titan's atmosphere. It measures the elements that are expected to exist, including carbon, nitrogen, hydrogen, oxygen, argon and neon, but also looks for new, unpredicted ones. Using this data, scientists hope to gain new insights into the chemical evolution on Titan and its possible relevance to the conditions on the early Earth.

GCMS is equipped with gas samplers that are filled early during the descent at high altitude, when the probe is still going very fast. The analysis of these samples is done later in the descent, when more time is available.

The Aerosol Collector and Pyrolyser is designed to collect aerosols, small droplets floating in the atmosphere, and subject them to a chemical composition analysis. It consists of a sampling device that extends from the probe, a pump to suck in the atmosphere, and a series of filters to capture the aerosols. The aerosols are then routed to NASA's GCMS instrument for analysis.

The Huygens Atmosphere Structure Instrument comprises various sensors for measuring the physical and electrical properties of the atmosphere, such as wind velocities, air temperatures and lightning discharges. It also incorporates a microphone that records sounds of the droning winds at Titan.

The Doppler Wind Experiment makes use of the radio signals that Huygens continuously sends to Cassini. The wind on Titan makes Huygens drift, inducing a measurable Doppler frequency shift in the signal.

The Descent Imager/Spectral Radiometer of the University of Arizona is an optical instrument with a number of different detectors. It is designed to shoot pictures of the surface of Titan and the clouds in the atmosphere, and to make spectrographic measurements to find out what they are made of.

At an altitude of 30 kilometers (19 miles) the thick orange haze opens up, and the imager can finally see the surface. Its images show mountains of clear ice, caught in a net of river channels filled with dark deposits. The dark material probably consists of complex organic molecules that the rain washes down the mountains. It also sees what looks like a coastline along a dried up lake.

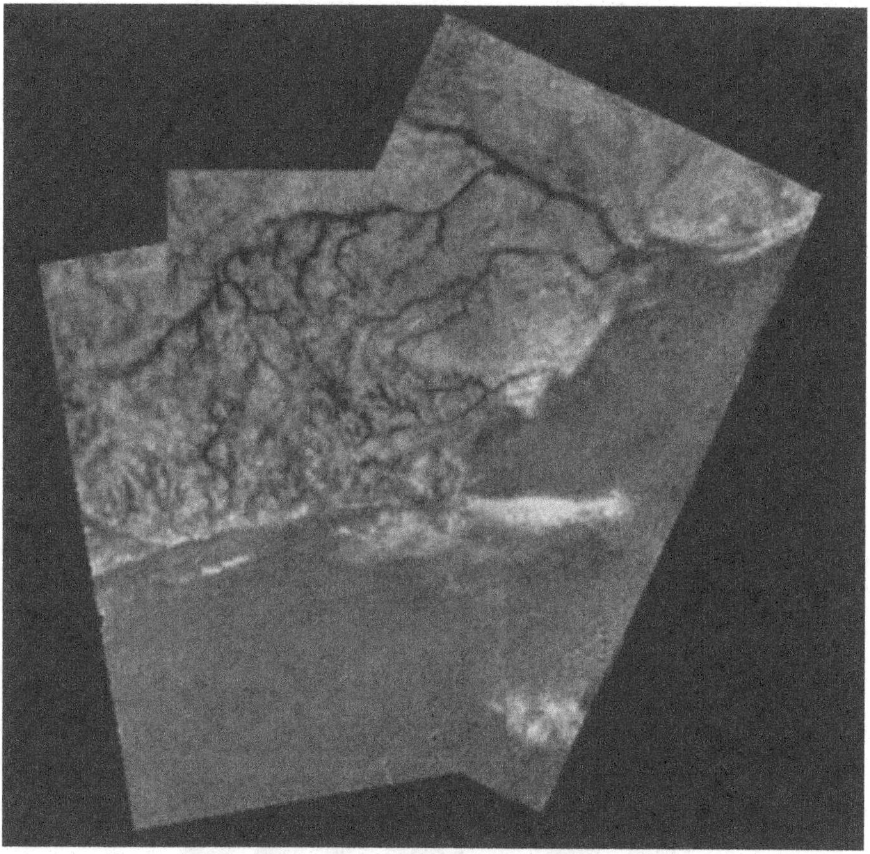

FIGURE 5.7 This image made by the Huygens probe from high altitude shows what looks like rivers full of dark material. [ESA]

Once Huygens is only a hundred meters or so above the surface, the instrument switches on a lamp to have a better look at the surface from nearby. The probe heads for a landing close to a border between the dark and light material.

Finally, Huygens hits the ground at 5 meters (16 feet) per second and the Surface Science Package can finally get into action. It has been specifically designed to find out what kind of surface Huygens lands on, in case the probe makes it that far (officially, Huygens was designed as an atmosphere investigator, not as a lander). The theoretical possibilities for the ground range from a rocky surface to methane-ice or snow, or even methane or ethane lakes and seas!

The package of sensors includes a so-called "penetrometer" to measure the impact deceleration (high for hard ground; lower for softer material), and equipment to determine all kinds of physical properties of the surface

FIGURE 5.8 The surface of Titan, as photographed by Huygens. The rounded blocks of ice in the foreground are about 10 centimeters (4 inches) in diameter. [ESA]

material, such as temperature, density, thermal conductivity, heat capacity and the local speed of sound.

Huygens hits a relatively soft layer of material, possibly some kind of ice, and sinks partly before settling down completely. Later the measurements will be compared with laboratory results, collected during tests in which the penetrometer was dropped in different buckets filled with all kinds of materials. The analyses will show that the landing spot probably consisted of icy grains with the consistency of sand.

Shortly after impacting the surface, the instruments measure an increase in methane gas. The heat of the active probe is warming up the ground, which apparently contains liquid methane that now evaporates.

Pictures of the surface show a vast flat plane, covered with rounded blocks of ice (probably water-ice) that have been rounded by the wind and, maybe, liquid erosion. It is a desolate place, but still it looks somewhat familiar. However, this landscape has not been shaped by water and air, but by liquid methane and an alien atmosphere! The orange veil of Titan has finally been lifted.

For at least another 2 hours Huygens continues to send images and measurements from the surface, but then Cassini loses sight of the little robotic explorer and decides to turn its large parabolic antenna back to Earth. Ground-based radio telescopes however continue to receive Huygens' signal until far beyond the expected lifetime of the batteries; they turn out to contain enough power to keep the probe alive on the surface for a total of 3.5 hours.

At the ground control center the scientists and engineers are eagerly awaiting the first Huygens data transmitted by Cassini. The moment at which the signals are expected comes and passes. Has something gone wrong? Then, 6 minutes later than anticipated, the computer screens start to fill with long rows of numbers; the probe's invaluable scientific data is finally pouring in.

However, not everything is well. Instead of two rows of data, only one is seen on the screens. As an insurance against failing equipment, Huygens sent its data via two separate transmitters working at different radio channels to Cassini. Now it appears that only one channel was received by the orbiter. Later the engineers find out that because of a command error Cassini was not tuned to receive the signals coming in on the other channel; Huygens was transmitting on two channels, but Cassini's receiver was only listening to one.

The data of most instruments has been distributed over the two channels and so half of all they collected has still been received over the single working channel. The transmitting of the descent pictures has for instance been alternating between the two channels, so even though half of all the images are lost, the scientists are still able to assembly a complete series showing the whole descent.

The most important data initially lost is that of the Doppler Wind Experiment, the data of which was only sent over the channel that Cassini was not picking up. Fortunately a network of radio telescopes on Earth has managed to receive some of Huygens' weak radio signals directly from Titan. Using data from large receiving dishes in Australia, China, Japan, the USA and Europe, the movements of the probe in the Titan atmosphere can later be puzzled together after all.

The Doppler data has actually been found to be extremely good; even the swinging motion of the probe under its parachutes is detectable!

The Huygens mission at Titan has only lasted for hours, but has given the scientist enough data to keep them busy for many years.

6

LAUNCH

THE launch is one of the most important moments in a space exploration project. It is a very short phase, as it takes roughly about 8 minutes to get into Earth orbit and only another couple of hours before the probe is boosted into interplanetary space.

The launch signifies the moment when a project moves from designing, building and testing (when the spacecraft can still be touched) into the operations phase, where it is gone forever. At that moment the project is handed over by those who helped it to grow to those who operate it in space.

ON THE ROAD

The launch preparations start with the transport of the spacecraft from the safety of the assembly and testing facility to the launch site. Sometimes spacecraft travel by plane, sometimes by ship and almost always by truck for at least a part of the journey. The means of transportation may even need to be taken into account during the design of the spacecraft. It can involve shocks and load directions that are very different from what the space probe can normally expect during launch and operation.

Engineers not familiar with the Russian launch base Baikonur are sometimes baffled by a strange spike showing up in the shock spectrum to which they have to design their spacecraft. It has nothing to do with the actual launch; it describes the shocks that may occur when the spacecraft is brought to the launch pad over a bumpy Russian railroad track.

Spacecraft are not designed for the dusty, windy and sometimes wet atmosphere on Earth, the planet on which they were created. They may be able to stand extreme heat, cold, radiation and rough landings on distant planets, but a bit of rain or condensation can easily destroy their delicate equipment. They are therefore meticulously packed in hermetically sealed containers, equipped with their own independent air conditioning.

At the launch site the containers are carefully opened inside a cleanroom. Some parts, such as antennas and solar arrays, may still need to be attached, and the whole spacecraft has to be checked one last time to ensure that it has survived transportation in mint condition.

SPACEPORTS

Interplanetary robots usually leave our planet from one of four major launch sites in the world.

The best known are the Kennedy Space Center and the adjoining Cape Canaveral Air Force Base in Florida. Apart from the famous launch pads for the Space Shuttle, they include many more facilities for the launching of all kinds of civilian and military rockets.

It is handy to launch your spacecraft from near the equator in an eastward direction, because it then gets the maximum speed benefit from the free velocity of the Earth's rotation (unless you want to launch a satellite to orbit over the Earth's poles). Spacecraft launched from close to the equator into lowly inclined orbits (i.e. at small angles with an imaginary line over the equator) get a free boost of 1,650 kilometers (1,030 miles) per hour. This saves an important amount of propellant, meaning that you can launch heavier spacecraft than elsewhere with the same launcher.

At latitude 28.5 degrees north, Kennedy Space Center is relatively close to the equator. Moreover, the vast expanse of the Atlantic Ocean lies to the east, so that spent rocket stages (and failing rockets) can safely fall into the sea.

Europe does not have an ocean to its east, only other relatively densely inhabited countries. As a result, launching interplanetary spacecraft from Europe is not safe.

Luckily France has French Guiana, an overseas department in South America. A launch center for Europe's Ariane rockets was constructed at the edge of the jungle there, near the fishing village of Kourou. As it is located even closer to the equator than Kennedy Space Center, rockets launched from Kourou get an even bigger swing from the Earth's rotation. Moreover, Kourou also borders the Atlantic.

Russia launches its interplanetary spacecraft from the vast area of Baikonur, also known as Tyuratam. In bygone years the Soviet Union was afraid that its primary launch base, including the nuclear rockets standing ready there, could be eliminated by a couple of US missiles if the pads were too close together. The cosmodrome therefore comprises about 50 launch pads that are spread out over an area greater than five times that of the city of Los Angeles.

To find the Soviet ICBM (Inter-Continental Ballistic Missile) test site, US intelligence employed its secret U-2 spy planes. These long-winged planes could fly higher than any Soviet plane or anti-aircraft missile of the time.

The Soviets tried to keep its enemies in the dark about the location of the launch base, but the U-2s started to follow major railroads in the hope that these would lead to the secret launch base. The idea was a success. In the summer of 1957, only weeks after the first Soviet ICBM test flight, a U-2 mission was able to shoot pictures of the R-7 rocket's launch pad in Tyuratam.

Even though the West now knew where to find the cosmodrome, the Soviets continued to try to misinform their adversaries about the location. When they registered their first human spaceflight by cosmonaut Yuri Gagarin in 1961 as an official world record, they noted Baikonur as the launch site. In reality, Baikonur is a mining town 350 kilometers (218 miles) downrange from Tyuratam.

Tyuratam is actually situated in Kazakhstan, which after the break-up of the Soviet Union became the independent Republic of Kazakhstan. As a result, the Russian base is now on foreign soil and the Russians have to pay rent to the Kazakhs to be able to continue their launches.

Japan launches its spacecraft from its Kagoshima Space Center, situated on the most southern of the country's four main islands. Because large fleets of tuna fishers make their living in this area, launches are only permitted during a number of days in January, February and August.

Apart from these launch bases, there are other existing sites where interplanetary spacecraft may start their voyages in the not too distant future.

China has three major launch centers, but has not yet launched any spacecraft beyond Earth orbit. However, with the flight of Yang Liwei on board his spaceship Shenzhou 5, and the two "taikonauts" on board Shenzhou 6, China has recently become the third country that is able to launch its own astronauts. It will undoubtedly also want to organize its own interplanetary missions, and has expressed intentions to launch robotic spacecraft to the Moon in the near future.

India has a launch base at Sriharikota Island in the southern state of Andhra Pradesh. The country has been launching satellites since 1980 and has announced plans to send a space probe to the Moon some time in the coming years.

A very special type of launch base is operated by Sea Launch, a partnership of American, Norwegian, Russian and Ukrainian companies. Their launch facility is not at a fixed location on land, but consists of a floating, self-propelled launch platform and a separate assembly and command ship.

As the launch platform – a modified oil rig – can be remotely controlled by the command ship's launch control center, no one needs to be on it during a launch. The command ship also houses a large assembly hall, where the Zenit-3SL rockets and payload are integrated before being transferred to the launch platform.

This unique concept means that the whole "base" can be moved to an ideal location exactly on the equator. Moreover, in the empty middle of the ocean there are fewer safety, security and scheduling constraints. Sea Launch mainly targets the launch market for commercial communications satellites, and until now has not launched any interplanetary spacecraft.

Spaceports do not only comprise rocket launch pads. They also include facilities for the preparation of the rockets and the spacecraft, and integration buildings for putting the spacecraft on top of the launchers.

Then there are propellant production facilities, specifically for the extremely cold liquid oxygen and hydrogen "cryogenic" propellants. These fluids need to be cooled, respectively, to –184 and –251 degrees Celsius (–299 and –420 degrees Fahrenheit) to keep liquid. A little bit is always evaporating, and this "boil off" process makes it difficult to transport them. If the travel distance was too long, all the propellant would evaporate before it could be put into the launcher.

Propellants for large solid propellant stages and boosters are also produced at the launch site, as it is much easier and safer to transport these items when they are empty.

Non-cryogenic, storable liquid propellants such as kerosene are usually produced elsewhere, then transported to the launch site and stored near the launch pads.

Launch sites also need mission control rooms for directing the launch and launch preparation activities, and a series of radar and optical tracking stations to monitor the rockets during flight.

Preparing for Launch

Planning of the launch campaign, i.e. everything directly related to the launch that has to be done before the spacecraft reaches orbit, is critical for the success of a mission. The launch preparations can take a long time; they have to be started at the right moment to ensure that the launcher will be ready for liftoff when needed.

The right moment for launch is determined by the orbit and position of the target planet, the orbit of the Earth, the Earth's rotation and the trajectory needed for the probe to reach its destination. Clearly, this can easily become very complicated, and the right situation for launch may not occur very often or for the correct length of time.

The available time period for launch, the so-called "launch window," can therefore be very small, and the interplanetary spacecraft has to be put on its way at a very precise moment. A launch at any other time would put the probe in a wrong orbit with respect to its target. For trips from Earth to Mars, the planets are in the right position for only a short period every 26 months.

For example, the launch window can have a duration of only a few minutes each day over a total period of a week. Missing the right moment on a certain day is no disaster, as another possibility will normally be available the following day. However, missing all daily launch opportunities for the overall possible launch window, such as a week in this example, means that the probe is no longer able to reach its target. The movements of the planets do not wait for us. Sometimes there may be no new series of launch opportunities for years, and perhaps never.

This last situation occurred with ESA's comet probe Rosetta, which missed its proper launch window because there were serious problems with the type of Ariane 5 rocket that was supposed to launch it (the Ariane launched just before Rosetta failed to reach orbit because of a design error in its new main engine). By the time the rocket system was back in operation, Rosetta could no longer reach its original target comet and a new comet in a suitable orbit had to be (and was) found.

The pre-flight operations for a spacecraft launcher begin with transporting all its elements to the launch base. For an Ariane 5 launch campaign in French Guiana, the rocket stages are brought over sea from Europe in one of two special transport ships, the *Toucan* or the *Colibri*. The trip takes 12 days.

After unloading at the Pariacabo dock up a sea inlet leading to the launch base, the stages are transported by road to the nearby spaceport. The Ariane 5 main stage, upper stage and the vehicle equipment bay with all the guidance, navigation and communication equipment are transferred to the Launcher Integration Building. The aerodynamic fairing, which protects the spacecraft before and during launch, is brought to the Final Assembly Building.

Inside the 58-meter-high (190-foot-high) Launcher Integration Building, the main stage is pulled up by a big crane and placed vertically on a mobile launch table. The two huge solid propellant boosters, having been filled with solid propellant in the Booster Preparation Building, also arrive in the integration building and are vertically docked to the main stage, one on each side. Next, the vehicle equipment bay and the upper stage are mounted on top of the main stage.

The nearly complete Ariane 5 is then checked to ensure that all elements have been connected securely and that all electrical connections are working.

Sixteen days later the doors of the Launcher Integration Building are opened and the rocket on its launch table is moved along its rail track to the Final Assembly Building, were it will be mated with its payload.

In the meantime the spacecraft in its storage container has arrived by plane at Rochambeau Airport in Cayenne, and driven to the Satellite Preparation Facilities of the base.

Here the space probe is unpacked inside one of the cleanrooms, separately shipped items such as solar arrays and large antennas are installed and the whole thing is checked to make sure that everything is in working order.

Most spacecraft have their own propulsion subsystem, which means that their tanks need to be filled before launch. As spacecraft propellant is often dangerously toxic, this takes a lot of preparation and needs to be done with great care.

Engineers responsible for filling the tanks need to wear special protection suits with helmets and supply lines that provide air from outside the cleanroom. It makes them look like astronauts.

After filling, the spacecraft tanks are put under pressure and the system is checked for leakages.

The spacecraft is then mounted on a Payload Adapter, which functions as the interface with the launcher. The Payload Adapter includes the mechanisms that will free the spacecraft from the launcher when it has arrived in orbit.

Once all this has been done the spacecraft is also transferred to the Final Assembly Building.

At the 90-meter-high (300-foot-high) Final Assembly Building the spacecraft on its Payload Adapter is hoisted on top of the Ariane 5 upper stage. Finally, the Payload Fairing is put over the spacecraft to structurally complete the launcher. The next step is to fill the upper stage and the rocket's altitude control systems with non-cryogenic propellant and to perform further checks on the entire combination.

Operations in the Final Assembly Building last about 14 days, after which the Ariane 5 with its payload is rolled out to the launch pad on rails, pulled by a powerful truck. The mobile launch table is then connected to the launch pad, propellant lines are hooked up to the main stage and electrical connections are made with the launcher and the spacecraft. The final countdown then begins.

Shortly before launch the tanks of the Ariane 5 main stage are filled with liquid oxygen and liquid hydrogen. Because these extremely cold fluids boil off rather quickly and are dangerous to work with, this cannot be done earlier. In the meantime, the launcher subsystems are checked and the computer program with the flight data for the automatic guidance system is loaded into the launcher's computer. The launch teams get ready in the control center and various tracking facilities.

Before launch, people have to be cleared from the area around the launch pad and airspace through which the vehicle will travel. Rocket launches are sometimes delayed because tourists come dangerously close in sports planes or boats. Dropped rocket stages or debris from an explosion could kill anyone who is too near the launch pad or is directly under the trajectory of the rocket.

At 6 minutes 30 seconds before liftoff, the countdown becomes automatic. The rocket itself is now working on its own, with all further actions being timed by its internal clock. From then on, apart from stopping the whole launch procedure, the people at the base can only watch and hope for the best.

Chapter 6

UP AND AWAY

The time for checking and adjusting is over. As the spacecraft sits on top of its launcher moments before liftoff, the minds of all those involved in its creation are spooked by nightmares of explosions, unknown mistakes that may have slipped by, critical issues that might have needed just a little more study. Soon years and years of work will be hurtling into space on top of hundreds of tons of burning explosives.

Launching spacecraft is still not routine, even though we have now been doing it for nearly 50 years. The pressures in the rocket engines, the velocities and the forces involved are tremendous, and yet rockets cannot be over-designed very much because that would make them too heavy to carry a reasonable mass into orbit. The margins for error are therefore rather low. On average, about two or three out of a hundred launches fail. This is an extremely high failure rate when compared to airplanes, for instance, where crashes are measured in numbers per millions of flights.

One minute before liftoff the electrical power supply from the launch pad is cut off and the rocket becomes autonomous, drawing power from its internal batteries. The countdown proceeds in French "... 3, 2, 1, Top."

Hydrogen and oxygen pour into the combustion chamber of the main stage's single Vulcain engine and are ignited. The hot gasses reach a temperature of 1,500 degrees Celsius (2,730 degrees Fahrenheit) and spew out of the nozzle, giving the launcher 115 metric tons (255,000 pounds) of thrust. However, this is not enough to push the 745-metric-ton (1.64-million-pound) Ariane 5 off the ground. For just over 7 seconds the launcher remains on the pad to check whether the Vulcain engine works properly.

Then the two powerful, 30-meter-high (98-foot-high) boosters are ignited and begin to deliver most of the thrust needed to take off. Once these unstoppable solid propellant rocket motors have been started it is no longer possible to prevent the Ariane 5 from leaving the pad. The combined thrust of the main engine and the two boosters at liftoff is about 1,200 metric tons (2.65 million pounds), the equivalent of the thrust produced by 11 Boeing 747 aircraft.

Majestically and with a thunderous roar the rocket climbs straight up and clears the launch tower. A set of small engines in the upper part of the rocket is used to roll the Ariane around its axis and correct its orientation. The big nozzles of the two boosters and smaller one of the Vulcain engine swivel to steer the launcher onto the right trajectory. Soaring into the tropical sky the Ariane 5 heads east over the ocean.

FIGURE 6.1 An Ariane 5 leaves the launch pad at Kourou in French Guiana. [ESA/CNES/Arianespace]

About 130 seconds after liftoff the rocket has already reached an altitude of 55 kilometers (34 miles). At that moment the two boosters have each spent their 238 metric tons (525,000 pounds) of solid propellant and become dead weight. Pyrotechnic devices free them from the rest of the launcher and separation rockets ignite to push the nearly empty steel cylinders away.

The now much lighter Ariane 5 continues its journey into space on the thrust of its Vulcain engine. The spent boosters fly on for about 100 kilometers (62 miles) and then, some 450 kilometers (280 miles) from the launch site, fall into the Atlantic Ocean.

Although the boosters can be recovered by ships for technical investigation, unlike the similar but larger boosters of the Space Shuttle, they are not reused. For Ariane 5 it was found to be less expensive to build new ones than to disassemble, clean, repair and reassemble old ones.

During flight through the atmosphere the payload fairing protects the precious spacecraft cargo against the pressure and heat of the compressed air flowing over the nose of the rocket. Inside, the fairing has a layer of acoustic absorption panels, to dampen the acoustic violence of the boosters from damaging the spacecraft.

Some 3 minutes into the flight, at an altitude of about 100 kilometers (62 miles), the Ariane 5 has left nearly all of the atmosphere behind. There is no longer a need for the more than 2-metric-ton (4,000-pound) heavy fairing, and it is jettisoned.

Two pyrotechnic systems are used to remove the fairing. Detonating the horizontal system ruptures the connection with the Ariane 5's upper part. The vertical separation system splits the fairing in two halves when it separates from the launcher.

Seven minutes later the first, main rocket stage runs empty and is jettisoned. The onboard computer ignites the second stage and the flight goes on with a much smaller configuration. Only the upper rocket stage, the Vehicle Equipment Bay and the spacecraft payload continue the ascent.

About half an hour after liftoff, depending on the target orbit, the spacecraft is finally released from what is left of the Ariane 5. The spacecraft is then on its own.

The launch phase is now over, and a hectic period starts for the people in the spacecraft control and operations center. They have to send the commands to deploy the solar arrays and antennas, and make sure that the probe is correctly oriented to enable them to communicate with it.

Next, the operations team has quickly to check that the space probe is fine and in the right orbit. This is the first time the spacecraft is really in space, and some simulations and tests done on Earth may prove to have been slightly inaccurate.

Parts of the probe may get warmer than predicted, which can be helped by reorienting the spacecraft to provide more shadow for the overheated parts. Some equipment may not have survived the launch 100 percent. This can be tested by looking at the telemetry data sent by the spacecraft and manual test operations of the troubled elements. The operations team may decide to switch over to backup equipment if the impact of malfunctions is serious, or adjust the mission plan to work around the problems.

New software with routine flight command sequences may be uploaded to replace onboard software that was only used just after launch. The programming of the onboard computer may need to be adapted to the differences between the real flight in space and what was foreseen before launch.

Once all equipment has been checked out, all immediate problems have been solved and the spacecraft has been put into a stable situation, routine flight operations begin.

Rockets

Most interplanetary space probes get launched on expendable "throw away" rockets. When the spacecraft is in orbit, nothing of these huge, expensive machines is left for reuse on other missions. The empty stages of these machines are dropped off along the way to get rid of dead weight, splashing into the ocean or burning up in the atmosphere.

The only partly reusable system ever employed for putting interplanetary spacecraft on their way is the Space Shuttle. The Space Shuttle, or Space Transportation System as it is formally named, consists of the Orbiter, a large External Tank and two enormous Solid Rocket Boosters.

The Shuttle Orbiter is the famous stubby plane in which the astronauts are launched and live while in orbit. It returns astronauts and material to Earth and is completely reusable, although an enormous effort is needed to refurbish the vehicle after each flight.

The External Tank is the large brown cylinder that carries most of the propellant for the three large rocket motors on the Shuttle Orbiter. The tank stays attached to the Orbiter for almost the entire ascent, but is discarded just before entering orbit. It falls back to Earth and burns up in the atmosphere because of its high velocity, which compresses the air

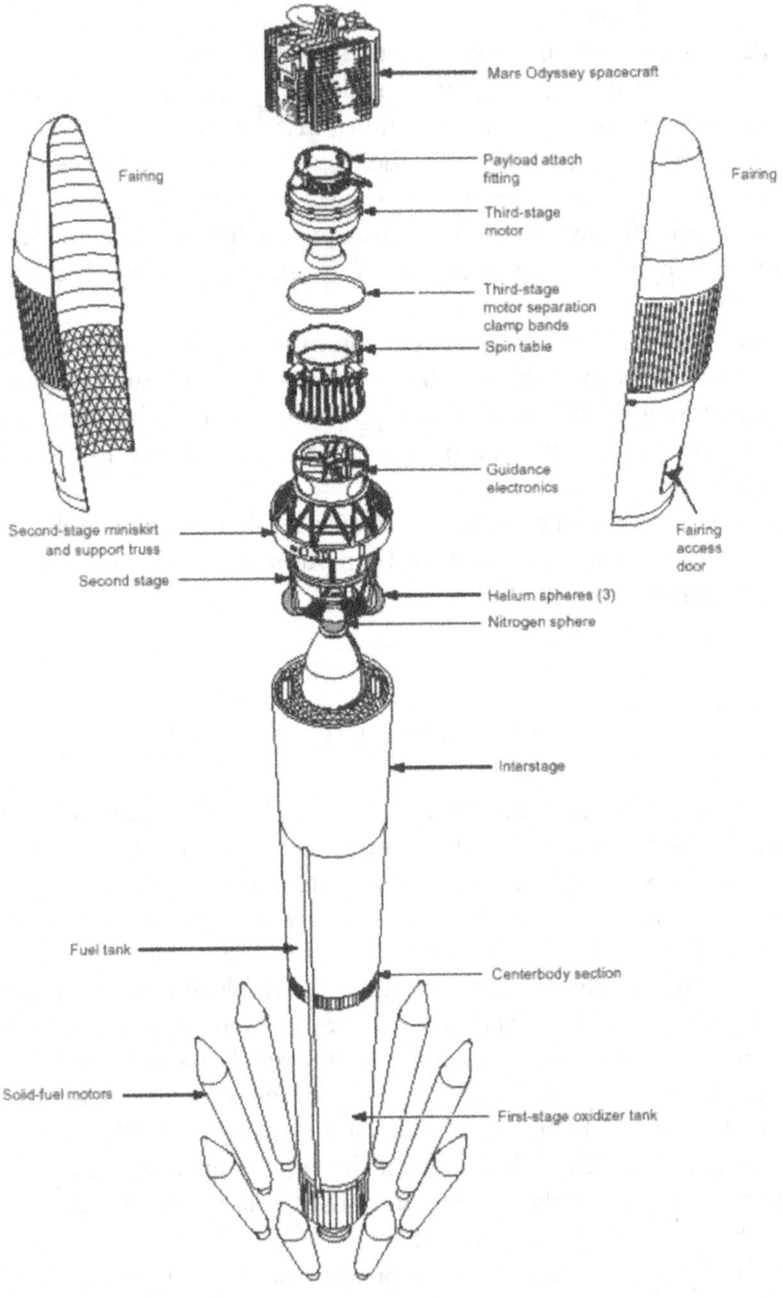

FIGURE 6.2 A Delta II launcher consists of many large elements that have to be assembled at the launch based. The spacecraft goes on top. [NASA]

resulting in extremely high temperatures (the effect is similar to a pump heating up when you inflate the tires of a bicycle, due to the compression of the air inside).

The two Solid Rocket Boosters attached to the External Tank are similar to those on Ariane 5, but much larger. They provide most of the thrust during the first 2 minutes of the Space Shuttle's flight. After burning out, they are ejected. However, unlike the External Tank they are not lost but are retrieved from the ocean after a slow descent by parachute. After a major refurbishment effort, they are reused and are later reloaded with new solid rocket propellant.

The reason that expendable rockets are still used more than reusable systems like the Space Shuttle, is that they are much less expensive to develop and much easier to operate. In fact, the maintenance of the Space Shuttle is so costly that, for the relatively few times per year that a satellite needs to be launched, it is cheaper to build and use an expendable, one-shot rocket.

Expendable launchers are inherently expensive in use: a medium sized launcher such as the Russian Soyuz Fregat can put a 1,100-kilogram (2,400-pound) spacecraft such as Mars Express on its way, but to do so it burns some 289 metric tons (640,000 pounds) of propellant and throws away 26 metric tons (57,000 pounds) of precious rocket hardware.

A Soyuz Fregat launch is relatively cheap in comparison to European and US rockets, costing in the order of $40 million. A US Delta II 7925 rocket can launch a 725-kilogram (1,598-pound) spacecraft such as Mars Odyssey out of Earth orbit for about $60 million. The Atlas V-401 that launched the 2,180-kilogram (4,810-pound) Mars Reconnaissance Orbiter in August 2005 has a price tag of about $90 million. An Ariane 5 launch of an even heavier space probe of 3,000 kilograms (6,600 pounds) such as the comet explorer Rosetta, costs over $150 million.

However, before the Columbia disaster, a Space Shuttle launch with all its complicated pre-launch activities and human spaceflight equipment for the astronauts was costing in the order of $300 to $500 million per flight. The additional safety constraints now put in place probably put the current costs well above half a billion. For launching spacecraft beyond a low Earth orbit it even needs an additional, and costly, expendable rocket stage.

Until we learn how to build and maintain efficient, cost-effective reusable launchers and have a need to fly them very often so that the average launch price drops dramatically, most interplanetary probes will continue to be launched with old-fashioned but effective expendable rockets. New versions of the Ariane, Atlas, Delta and Soyuz rockets are still being developed, and it doesn't look as if they will become obsolete very soon.

Chapter 6

ORBITS

Do you remember the earlier explanation about how a large gun on top of an extremely high tower could theoretically put an object into orbit? At the right speed a satellite will not fall back down, but will continuously circle the Earth.

At a higher velocity of about 12 kilometers (7.5 miles) per hour, the Earth's gravity will even lose its grip and the spacecraft will be able to break out of orbit and escape into outer space. At that point it becomes a satellite of the Sun, just like the planets.

If the spacecraft's orbit is highly elliptical and therefore crosses the more or less circular orbits of the planets, it may possibly be able to travel to other worlds.

Moving from Earth to another planet is very different from flying from one city to another. Cities don't move and, moreover, are always located on the surface of the Earth. That makes finding your way a relatively simple two-dimensional problem, and a flat map is sufficient for navigation. However, planets are constantly in motion with reference to everything else; they circle the Sun at different speeds and seldom exactly in the same plane.

Moreover, once a space probe is on its way, Newton is in the driving seat. The orbital mechanics that he, for the first time, described mathematically make the probe follow a curved, elliptical trajectory. This orbit can be changed only slightly because the amount of rocket propellant available is limited.

So, before leaving Earth we have to make sure that the target planet is at the right location when our spacecraft crosses its orbit. The departure time or "launch window" is determined by this. We also need to configure our trajectory in such a way that the probe will rendezvous with the planet at the right speed – not too fast if we want to go in orbit around it, and not too slow if we only want to fly by and continue our journey. Aiming straight at the planet all the way is not a good idea if you don't want to crash into its surface.

The most energy efficient way of traveling from one planet to another – the one that requires the lowest speed and therefore the least amount of propellant – is called a Hohmann transfer. In a Hohmann transfer, a spacecraft follows an elliptical orbit that just touches both the orbit of the Earth and that of the target planet.

To reach a planet further from the Sun, the probe will depart Earth at the point on its elliptical Hohmann transfer orbit that is closest to the Sun (known as the "perihelion"), and will meet the target planet at the transfer

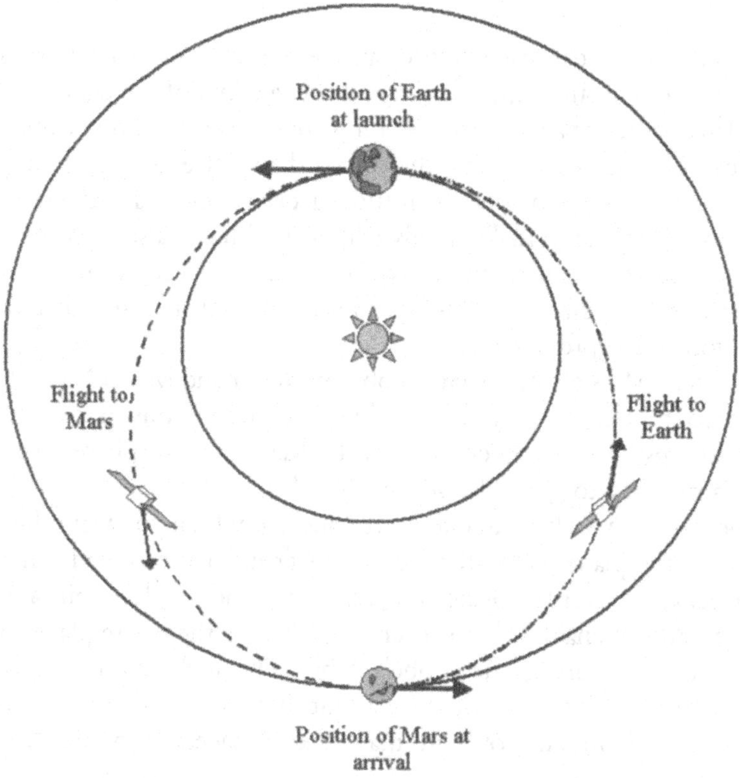

FIGURE 6.3 A space probe from Earth using a Hohmann transfer to reach Mars, and one launched from Mars using a Hohmann transfer to get to Earth.

orbit's furthest point (known as the "aphelion"). To reach planets closer to the Sun, the process is reversed. The spacecraft will thus travel exactly half of an ellipse to get to its target.

For a flight to Mercury for instance, a probe launched from Earth starts at the aphelion of its Hohmann trajectory. To change its orbit so that its perihelion gets closer to the Sun and touches the orbit of Mercury, the probe needs to decrease its orbital energy. It does that by accelerating opposite to the direction of the Earth's motion around the Sun. It effectively brakes and thereby "falls" toward the Sun.

To get to Mars, a spacecraft departs at perihelion, and thus has to increase its orbital energy by accelerating in the direction of the Earth's motion. That will make its aphelion equal to the distance of Mars's orbit.

For flying to Mars, the best launch position is some months before Earth is about to overtake Mars on the "inside track" in their concentric orbits around the Sun. This ensures that by the time the spacecraft arrives at the orbit of Mars, the red planet will also be at that point in its orbit.

When the probe has reached its destination, it may also have to brake to get into an orbit around it, depending on the velocity of the planet and the speed at which the spacecraft reached it. Once there, the orbit around the planet can be adjusted by using the onboard rocket engines. Braking has the effect of lowering the orbit's altitude; accelerating will make the orbit higher. In this way an initially highly elliptical orbit can also be made more circular, which is often preferred for planet observation missions. (However, a fully circular orbit is seldom achieved, because that would require too much propellant.)

The recent Mars orbiters Mars Global Surveyor and Mars Odyssey have used "aerobraking" to slow down. This technique consists of carefully calculated dips into the upper layers of the Martian atmosphere, using the aerodynamic drag to gradually adjust the orbit.

Before arrival, you have to choose the angle at which you want the orbit to intersect the planet's equator, i.e. the orbit inclination. You can make the spacecraft fly over the planet's poles, for instance. Polar orbits are very handy, because it enables the spacecraft to observe the entire planet while it slowly revolves under the probe's orbit. The inclination can also be altered once in orbit around the planet, but that will use a lot of propellant (and carrying too much propellant may make a spacecraft too heavy for its designated launcher).

Therefore, being left behind on Earth, how do we keep track of where an interplanetary spacecraft is and how fast it is moving? This is important, as along the way we will need to fine-tune the navigation and observation commands for the spacecraft when, for example, it needs to know when to fire its rocket engine and where to point its instruments.

The direction of the probe is easy to establish: it is where its radio signals are coming from. But to find out how far away it is requires some special measures.

To range the distance from Earth, we can place specially coded signals (called ranging tones) on the radio messages sent to the spacecraft. When they are received by the probe, it immediately puts them on its returning radio signal. By noting exactly the times at which the ranging tones left Earth and when they were received, and taking into account the speed of the signals (the speed of light), the round-trip distance can be computed.

In reality it is not as simple as that, of course. For really accurate measurements, we also need to know how long it took the ranging tones to "turn around" inside the spacecraft's electronics. Similarly, we should know how long it took the signals to get from the ground control computer to the transmitting ground station antenna. We also need to know how far the Earth moved while the ranging pulses were traveling to the spacecraft.

The minute delay within the spacecraft's communications equipment can be measured during pre-launch testing, as can the delays in the ground equipment. We can compute the movement of the Earth from detailed astronomical observations. Taking all this into account, we are currently able to measure the distance to a spacecraft far out in the Solar System within an accuracy of a few meters!

The velocity of the spacecraft can be determined by measuring the Doppler frequency shift in the radio signals it transmits, as described previously for Messenger and Huygens.

Once the spacecraft arrives near its target, we can also use the background stars on the pictures it sends to determine its position and velocity. It works similarly to how, on the old sailing ships, people determined their position by celestial observation. This so-called "optical navigation" can even give a more precise analysis of the spacecraft's trajectory than ranging and the Doppler effect alone.

7

DISTANT DESTINATIONS

IN this chapter we will have a look at how the diversity of robotic space probe missions – and, with them, our knowledge of the Solar System – have dramatically expanded over the last 20 years or so. Each mission targeting a planet, moon, comet or asteroid presented us with its own challenges. Each of them also rewarded humanity with unique scientific rewards when successfully completed.

Learning by trial and error, and often not solely driven by curiosity but also significantly by political and industrial competition, humans probed further and further into the Solar System – most often not in person, but by virtual presence with robotic explorers.

The history of space exploration is a bit like a video game: only when you have mastered the tricks required to complete a certain level can you go to the next, and each subsequent level presents you with new, even more demanding challenges. If you fail to reach the end of a level, you have to repeat it all over again unless you have more than one "life" (in our case, a backup spacecraft).

After we managed to put a satellite into Earth orbit, the next obvious step was to reach our own Moon – a very convenient target for learning the trades of robotic exploration without going to far from home. As the Moon is in an orbit around the Earth, it is continuously accessible and can always be found in an easily predictable location. The Moon is nearby, so it can be reached in only a matter of days. All this makes it possible for successive probes, each one improved on the basis of previous missions, to be launched to the Moon one after another with little time in between.

As conditions near the Moon are similar to those in Earth orbit, lunar orbiters could initially be very similar to Earth-orbiting satellites (in fact, the Moon is itself an Earth-orbiting satellite). Furthermore, the Moon has no atmosphere, so clean, sharp observations from orbit are relatively easy. Also, landing on the Moon is not too difficult, owing to the low gravity (only one-sixth of the gravity on Earth), lack of wind and weather, and the fact that on a small scale the lunar surface is rather smooth. There are not too many large rocks or steep crater walls that could make a lander topple over.

As the Moon always shows the same side to us, it was easy to stay in contact with probes landing on its nearside. Also, its proximity results in small communication delays between ground stations and lunar spacecraft of only a few seconds.

The next planetary destinations, Venus and Mars, presented an entirely new range of complex challenges. As each of these planets follows its own elliptical orbit around the Sun, independent of the Earth, probes sent to investigate them need to be launched within very specific timeframes.

The distance to these planets means that, seen from the Earth, they present a much smaller target in space than the Moon. They are thus more likely to be missed if our orbital calculations are not precise enough. The problems increase for the planets beyond Mars or closer to the Sun than Venus.

If the Sun were the size of a grapefruit, Earth would be the size of a grain of sand orbiting at a distance of 10.5 meters (34 feet). Venus would be another grain of sand orbiting closer to the Sun at 70 percent the Sun–Earth distance, and Mars would be an even smaller particle going around at a distance of 16 meters (52 feet). The next planet, giant Jupiter, would be a small marble at 54.5 meters (178 feet) and the outermost of the original nine large planets, Pluto, a speck of dust as far off as 414 meters (1,359 feet) from the Sun.

Try to imagine our microscopically tiny space probes traveling between these small objects, all within an area the size of a large parking lot. Looking at it in this way, it seems a miracle that we are ever able to rendezvous with any planet at all!

FIGURE 7.1 All the major planets of the solar system except tiny Pluto have now been imaged by space probes. From the top: Mercury, Venus, Earth and the Moon, Mars, Jupiter, Saturn, Uranus, Neptune. [NASA]

For spacecraft, life is easier near Earth (and this is also true for living organisms). The further away from it, the more difficult it gets. Traveling near Mars, which is much further from the Sun than the Earth or Moon, it is much colder and there is less light for solar arrays to convert into electrical power. Going near Venus, which is much closer to the Sun, spacecraft run the risk of overheating. Conditions near Mercury, even closer to the Sun than Venus, are extremely hot and radiation levels are lethal.

Nevertheless, as the next paragraphs will show, our sturdy robotic science explorers have now ventured nearly throughout the entire Solar System.

TO THE MOON

The Moon orbits quite close to us, on average at a distance of only 384,000 kilometers (239,000 miles). For centuries its surface has been studied through increasingly large and powerful telescopes. Nevertheless, until we were able to send spacecraft to inspect it, we knew very little about our neighbor in space.

Stuck on Earth, astronomers could see that the Moon was quite barren and had no visible vegetation or water on its surface. The areas that were named "seas" and "oceans" in classical times were actually found to be dark, relatively smooth plains of solidified lava.

Nor did it seem to have an atmosphere of any appreciable thickness as the horizon can always be clearly seen; the border between the sunlit side and the night side always appeared sharp and direct; and no clouds were ever observed to drift over the lunar surface. Moreover, the surface never seemed to be disturbed by wind, weather or anything else.

We could see lots of craters, small and large, but it was not clear whether they were volcanic in nature or the result of a heavy meteorite bombardment. There were also lots of mountains and winding valleys.

The first visits

In January 1959 Earth sent its first envoy to another world. The Soviet Luna 1 was intended to hit the Moon, but due to the primitive guidance methods of the time the probe actually flew past it, missing its target by just under 6,000 kilometers (3,700 miles). Typically, the Soviets claimed that they had intended to zoom past the Moon. Nevertheless, it was a great achievement; never before had humanity gotten so close to another world.

With Luna 2, following shortly in September of the same year, the Soviets managed to hit the lunar surface and thus placed, or rather smacked, the first man-made object on the Moon.

FIGURE 7.2 In 1959 the Soviet Luna 2 became the first spacecraft to hit the Moon. [Russian Space Agency]

Luna 3, launched a month later, went around the Moon and showed us, for the first time, images of the lunar farside. Not surprisingly, it proved to be as inhospitable as the nearside that is always facing Earth, but there were fewer dark "seas" to be seen.

Although spacecraft imaging techniques were still crude and the pictures therefore rather vague and blurry, the Luna 3 images nevertheless satisfied the curiosity of a Frenchman who had promised a crate of the best Champagne to whoever would show him the back of the Moon. New Year found the Russian Luna 3 technicians enjoying some well-deserved bottles.

In the meantime the USA had only managed to react with a single, more or less successful, Moon probe, Pioneer 4, in March 1959. It didn't go into lunar orbit, it didn't get closer than about 60,000 kilometers (37,000 miles) to the Moon, and it didn't even carry a camera. That was clearly no match for the string of Soviet spectaculars.

America was also behind in human space missions. The Soviets had been the first to send a man into Earth orbit, the first to send a woman in space, and the first to make a spacewalk outside the relative safety of a spaceship. This prompted president Kennedy to make his historic announcement in May 1961 to land a man on the Moon before the end of the decade. In his famous speech to Congress, he called the space race a battle between freedom and tyranny. The old Moon had become the subject of an ideological and technical race. For billions of years it had looked down on the Earth and recently watched mankind taking over the Earth. Now it was to become part of human evolution itself.

Preparing for Apollo

If it was to land people on the Moon and start developing the necessary spacecraft, America very quickly needed to discover what to expect on the lunar surface. Before Kennedy's announcement, NASA had decided to launch a series of kamikaze space probes to crash on the Moon and make detailed pictures of the surface while going down. Success with these Ranger missions would not come easy, however.

The Moon does not orbit in the same plane as the Earth's equator. Because of the 23.5-degree tilt of the Earth's axis with respect to the plane of its orbit, and the fact that the Moon's orbit is inclined by about 5 degrees to the plane of the Earth's orbit, the Moon's orbit may be inclined up to 28.6 degrees to the Earth's equator.

Because of this, the ideal moment to launch a probe to the Moon only occurs once every two weeks, approximately. It is, of course, rather easy to miss this launch appointment, for instance due to a problem with the rocket or because of bad weather.

To create more opportunities, i.e. to widen the launch window, a spacecraft can first be launched into a so-called parking orbit – a low orbit around the Earth that is parallel to that of the Moon. There, the probe simply orbits around until it is in the right position to continue its journey to the Moon. It will meet this position once on every orbit, about every 90 minutes, which is much better than the twice-per-month launch window back on Earth! When the right moment arrives, a further boost by the last stage of the launcher will put the probe on its way.

The Ranger spacecraft also followed this parking orbit method. The first two were initially launched successfully, but once in low Earth orbit the rocket stage that was to send them on their way to the Moon malfunctioned.

The third Ranger experienced guidance errors that sent it flying far past the Moon, missing it by 37,000 kilometers (23,000 miles). Ranger 4 actually managed to hit the lunar surface, but almost nothing on the probe worked and no useful data was returned. Ranger 5's solar panels shorted out about 75 minutes into the mission, and with depleted batteries and thus completely inactive it flew past the Moon uselessly at a distance of 720 kilometers (450 miles).

Confronted with the humiliating series of failures, NASA decided to severely simplify the over-engineered design used for the Rangers up to that point. All payload instruments not strictly necessary were removed, and only the essential battery of six cameras remained (two cameras, a wide-angle and a narrow-angle imager, looked down at the entire surface under the probe, while the other four each covered a quarter of the view).

The first of the new breed, Ranger 6, was launched in January 1964. Initially going the wrong way, NASA operators managed to have the probe make a successful course correction that aimed it straight for the Moon. With only some 2,000 kilometers (1,240 miles) to go, the command was sent to turn on the cameras. Nothing happened. Ranger 6 fell to its doom without returning a single image.

Success was finally achieved on July 31, 1964, when Ranger 7 came dashing down to the lunar surface with its cameras taking in a view never seen before. The pictures sent back to Earth revealed details down to 1 meter (3 feet) in diameter. This was a thousand times better than had ever been achieved with a telescope on Earth.

More than six months later, in February 1965, Ranger 8 was launched and returned more pictures of the Moon. Encouraged by the successes, NASA decided to broadcast the pictures of Ranger 9 live over television as it approached the surface in March 1965. Viewers of all major US television channels watched the view of the crater Alphonsus grow bigger and bigger on their screens, until the probe hit the surface. Mass media had reached the Moon.

In the meantime it was the turn of the Soviets to cope with failure after failure. None of their lunar probes launched after Luna 3 had delivered anything useful; the only partial success was Luna 4, which got into the wrong lunar orbit and only delivered some measurements on the solar wind and cosmic radiation. (Luna probes attaining Earth orbit but failing to reach the Moon were officially listed as Cosmos satellites to hide their failure; Soviet launches that did not even make it into orbit were not disclosed.)

After hard-impact kamikaze missions, the next step was to touch down softly, to have a better look at the lunar surface and to prove that it could be done. This meant that probes had to be equipped with rocket-braking systems capable of bringing down the velocity to virtually zero just before touching the surface (without an atmosphere, parachutes could not be used).

The first four of a new series of Soviet soft-landing Luna explorers crashed on the Moon (Lunas 5, 7 and 8) or missed it entirely (Luna 6), but Luna 9 was a big success. On February 3, 1966, it sent back three panoramic black and white pictures from the lunar surface. A British radio astronomer managed to pick up the non-encoded television signals sent to Earth by the Luna, and to the annoyance of the Soviets the pictures appeared in a British newspaper before they had the chance to officially present them to the world.

The pictures showed that the Luna 9 landing site in the Ocean of Storms was a dusty, rather flat area covered with rough rocks of all sizes. Most importantly, the surface proved to be able to support a heavy spacecraft; before the landing some scientists had predicted that anything or anyone attempting to land on the Moon would sink and disappear in meters of lunar dust.

Luna 10 was not supposed to land, but to become the first ever satellite in a stable orbit around the Moon. It did this very successfully and thus proved that it could be done. It also measured the amount of radiation emitted by radioactive materials on the Moon, and showed this to be far below danger levels for future human explorers.

Soon after the USA also managed to land gently on the Moon with its Surveyor spacecraft. More sophisticated than Luna 9, they not only made pictures of the surface, but also scooped up samples to test the thickness and consistency of the gray dust covering the surface. Of the seven Surveyors launched between May 1966 and January 1968, only the second and fourth were unsuccessful.

In the meantime the USA also put five Lunar Orbiter spacecraft in orbits around the Moon, between August 1966 and August 1968. These mapped almost the entire lunar surface to find the best landing sites for the future Apollo crewed moon missions. Lunar Orbiter 1 also made the first pictures of the Earth as seen from the Moon.

Now everything was set to send crewmembers. The Soviets managed to make the first return flight to the Moon and back with their unmanned Zond 5 spacecraft in September 1968. Zond 5 even carried a "crew" of plants, seeds, insects and tortoises. However, only three months later America put the crew of Apollo 8 in orbit around the Moon and returned them safely back to Earth.

Two more Apollo test flights followed, and on July 20, 1969, Apollo 11 astronauts Neil Armstrong and Buzz Aldrin became the first people to walk on the surface of the Moon. The Soviet crewed program was by that time in complete disarray; there were problems making their spacecraft work and during each test flight the giant N-1 moonrocket blew up shortly after leaving the launch pad.

In a last act of desperation, during the Apollo 11 mission the Soviets tried to send their Luna 15 lander to the Moon and have it return with a few surface samples. The last chance to steal away some of the glory from the American Apollo program died with Luna 15's crash on the lunar surface. Later, in September 1970, Luna 16 managed to launch back a handful of lunar dust to Earth, but by then the Apollo astronauts had already returned bags full of soil and rocks.

Robot rovers

America went on with its crewed lunar missions with Apollos 12 to 17, then left the Moon alone for over 20 years. The Moon Race had been won, and the American people and their politicians had lost interest. NASA now focused on the development and operation of the Space Shuttle and a large space station (which eventually became the International Space Station currently in orbit).

The Soviets, however, continued with their robotic lunar missions, partly to show that they did not need expensive crewed landings to do science on the Moon. (Their entire failed crewed Moon program was not divulged until the collapse of the Soviet Union, although CIA spy satellites had kept the US military up to date on the N-1 moonrocket launch attempts.)

In November 1970 they landed Luna 17, carrying the Lunokhod 1, a robotic vehicle the size of a golf cart that was remotely controlled from Earth. It had eight wheels with electric motors and was steered by rotating the wheels on each side at different speeds (much like a tank on Earth). A hinged cover with solar cells was opened during the day to provide electrical power, and closed at night to keep the machine warm. It came down on a separate landing module, equipped with extendable ramps to enable the vehicle to drive off onto the surface.

For 11 months Lunokhod 1 strolled over the lonely surface, making 20,000 pictures. In 1973 it was followed by Lunokhod 2. Although less spectacular than the crewed Apollo missions, the Lunokhods proved the technology for a new breed of robotic space explorer: the planetary rover.

FIGURE 7.3 The Soviet Lunokhod lunar rovers were remotely controlled by operators on Earth. It had a lid with solar cells that could be closed at night to prevent the rover from getting too cold. [Russian Space Agency]

The last Russian lunar mission of the twentieth century landed on the Moon on August 18, 1976. Luna 24 drilled 2 meters into the lunar soil and returned these valuable samples to scientists on Earth a few days later.

Modern missions

After the massive assaults of the 1960s and early 1970s it was fairly quiet on the Moon. In 1990, while on its way to Jupiter, the Galileo orbiter made the first good pictures of the Moon's south pole. However, in 1994 the Moon received some further attention when in February of that year Clementine arrived to orbit around it.

Clementine was a joint project of NASA and the US Department of Defense. It was designed to investigate the Moon using sensor technology that would also be useful for new military satellites such as those of the SDI "Star Wars" rocket shield project. The probe was built in record time and by a small team at relatively little cost.

For the first time ever, Clementine mapped the entire Moon in visible light, ultraviolet and infrared. It made a spectacular discovery when it beamed radar signals into the dark craters of the lunar south pole: the radio beams bounced back with twice the normal signal strength for the Moon's surface. Such high reflectivity suggested the presence of ice in these deep craters that are in permanent shadow from the warmth of the Sun!

Maybe this ice, if it was really there, had accumulated by impacts of comets consisting of water-ice, never evaporating because of the eternal low temperatures deep inside the south pole craters. These and other intriguing questions triggered a new NASA mission that was launched in 1998: Lunar Prospector.

Lunar Prospector did not have a camera on board, as it would not be able to see anything inside the dark craters in any case. Instead, it carried a suite of spectrometers, a magnetometer and an electron reflectometer to investigate the composition of the lunar soil from orbit. It was hoped that the instruments would find proof of the existence of water-ice on the Moon. Lunar Prospector did find traces of hydrogen in the south pole craters, but the jury is still out on whether this means that there is water-ice or not.

At the end of its mission, Lunar Prospector was even crashed into a deep crater on the south pole of the Moon, in an attempt to throw up a cloud of icy material. It was hoped that this could be analyzed by ground-based telescopes, but nothing was seen.

If there really are large quantities of water-ice to be found on the Moon, this may be very valuable for future human exploration. The water can be used for drinking and cleaning, and could be split into oxygen and hydrogen to be used as rocket propellant. Moreover, the oxygen could be used in the life-support systems of crewed lunar bases and other spacecraft.

The latest mission to the Moon is ESA's SMART 1, the ion-propelled spacecraft presented earlier in this book. At the time of writing, the little explorer is still actively mapping the surface composition of the Moon. SMART 1 is also looking for water-ice inside the south pole craters.

A new Moon

What has all this exploration of the Moon taught us? Quite a lot in fact, and much more than we could ever have hoped to find out using only telescopes on Earth.

The Moon is quite large compared to Earth, about 27 percent of Earth's diameter. All other moons in the Solar System, although sometimes larger in actual size than our Moon, are much smaller in comparison to their mother planet. Many scientists therefore regard the Earth–Moon system as a "double planet." However, the Moon is very different from Earth: it does not have an atmosphere, not much water, if any, and certainly no life.

The Moon has always played an important role in our lives, especially in the past. Its light made the night less dark and the phases of the Moon formed a way to measure time and maintain a calendar. The lunar phases also helped early seafarers to keep track of the tides, which was important for fishing and entering or leaving harbors.

Its mysterious motions across the sky, the continuous changing of its appearance and the terrifying phenomena of lunar and solar eclipses that the Moon causes suggested that it was associated with the gods. Even now, after people have walked over its surface, the Moon still fills us with a sense of mystery and romance and triggers people to change into werewolves even in modern movies.

Long before the dawn of humans the Moon had played a crucial role in the formation of life. It is currently believed that the tidal pools that were sometimes under water, and sometimes dry due to the gravitational attraction of the Moon on the sea, are the places were life first started. Moreover, as it circled the Earth, the Moon has had a stabilizing effect on the tilt of the Earth's axis, helping to keep the climate relatively stable.

The robotic and human missions to the Moon have in half a century taught us much more about it than all the previous thousands of years of astronomical observations. Thanks to space exploration, our knowledge of the Moon has vastly increased within the span of only a few decades. As Earth and Moon have shared 4,500 million years of history together, retracing its past helps scientists to better understand what has happened to our own home planet.

One of the main questions about the Moon is: Where did it come from? The most widely accepted theory is that when the Earth was perhaps 50 to 100 million years old, some 4.6 billions of years ago, it collided with a massive, wandering planetary object the size of Mars. The impact was a glancing blow, and the debris that was knocked off our planet and the perpetrator during this enormous crash went into orbit. Over time, this material merged to form the Moon.

The impact would have completely disturbed the outer layers of the Earth, so the theory also holds important implications for how we think the Earth has formed. In fact, if the outer skin of the Earth indeed was blown off, then the Moon should mostly contain relatively light material; the heavier items like iron had at the time already sunk to the core of the Earth, so the outer parts that formed the Moon should contain mostly lighter materials such as magnesium and aluminum.

Whether the Moon contains more light materials than Earth is something that is currently being investigated by the SMART 1 orbiter, and its results have the potential to make or break the theory about the formation of the Moon.

After the Moon was formed and its outer layers solidified, it was bombarded by rogue meteoroids, comets and asteroids that scarred its surface with innumerable craters of all sizes. The very largest impacts cracked its outer shell, allowing liquid magma to flow over the surface and form the relatively smooth lunar "seas."

The damage was never cleaned up by the erosion of water, air and plate tectonics as happened on Earth, but stayed visible forever. As a result, when we look at the Moon in the night sky we can still get an idea of the horrible ordeal it has suffered over billions of years.

Now the Moon seems silent and dead. However, sometimes astronomers (often amateurs, as they are about the only group still studying the Moon through telescopes) report having seen strange light effects or hazy gas clouds on its surface. The Moon may not be as geologically dead as we think, and some volcanic activity may still be happening on a very small scale.

UNROMANTIC VENUS

As the brightest object in the sky after the Sun and the Moon, the ancient Greeks named Venus after their goddess of beauty. Even in recent times Venus had a bit of a romantic lure about it, as some scientists suspected the surface to resemble a kind of prehistoric version of Earth. After all, Venus is only slightly smaller than Earth and the clouds that could be seen through telescopes clearly indicated a thick atmosphere with possibly lots of water. Its closeness to the Sun would mean that the average temperatures would be higher, so maybe it was hiding tropical forests teeming with life beneath its veil of dense clouds.

Greenhouse nightmare

Then our robotic explorers unmasked it for the hell it actually is; an extremely hot world of rolling plains, resurfaced by volcanic activity, crushed under a very heavy carbon dioxide atmosphere with a surface pressure 90 times that on Earth. That is as much pressure as a deep ocean submarine is exposed to at about 900 meters (3,000 feet) depth on our planet. Dark brown clouds drop "rain" of concentrated sulfuric acid. Temperatures on Venus reach 470 degrees Celsius (880 degrees Fahrenheit), which is hot enough to melt lead. Because of this, metals vaporize and condense at cooler, higher elevations; the mountains and hills may actually be coated by condensing lead vapor!

These appalling conditions are the result of a disastrous greenhouse effect. Heat from the Sun, already more intense than on Earth because Venus stands closer to it, is trapped by the thick atmosphere that acts as a planet-covering blanket.

On Earth we are concerned that our industries and cars may be putting so much carbon dioxide into the atmosphere that a greenhouse effect has been started. Some scientists are afraid that this could raise the average temperatures on Earth by several degrees, resulting in melting polar caps, floods and severe changes in global weather.

On Venus the greenhouse effect is caused by the huge amounts of carbon dioxide in the atmosphere, which were probably released in the past by volcanoes. Studying Venus can tell us a lot about the actions of greenhouse effects and what they can do to a planet. This directly links interplanetary space exploration with our trying to understand the greenhouse effect that could be happening on Earth, and as a consequence links it to our very lives.

Into the unknown

In 1960 the USA launched the first probe that traveled all the way to the orbit of Venus, although that planet was not in the right orbital position for an encounter. The Pioneer 5 mission was not intended to visit any planet, but merely to test communication with space probes at large distances.

Earlier spacecraft had barely reached beyond the Moon, so staying in radio contact with a probe at the distance of Venus was no small feat. Pioneer 5 demonstrated that it could be done, and also sent us valuable information about the Sun's magnetic field and the solar wind.

The Soviet Venera 1 was the first spacecraft to fly by Venus, when it passed the planet within 100,000 kilometers (62,000 miles) in May 1961. However, unfortunately contact with the spacecraft had already been lost three months before, so the mission was not a scientific success.

The first spacecraft to return data about Venus was NASA's Mariner 2, which was a backup for the Mariner 1 mission that failed shortly after launch. However, its success did not come easily.

On its way to Venus the spacecraft suddenly lost its attitude control for about 3 minutes, perhaps because of a collision with a small object. Later the energy output from one solar panel suddenly dropped, and the science instruments that were active during the cruise flight to Venus were turned off. A week later the solar panel started to work properly again for some time, but then suddenly failed completely. Fortunately Mariner 2 was by then close enough to the Sun to be able to operate normally on one panel only.

In December 1962, Mariner 2 passed the planet at a closest distance of 34,800 kilometers (21,600 miles), and ended up in an orbit around the Sun.

The discoveries made by Mariner 2 include the high surface temperatures and pressures, the carbon dioxide atmosphere, the continuous cloud cover with a top altitude of about 60 kilometers (37 miles), the lack of a detectable magnetic field, the slow rotation rate of Venus, and that fact that it rotates around its axis in the opposite direction to the Earth and all other large planets.

In the meantime Russia had to cope with a whole string of unsuccessful Venus missions, which failed either due to problems with the launcher or because contact was lost before the probes reached their destination. However, in October 1967 Venera 4 descended through Venus's atmosphere, transmitting data during the descent. It gave us a first idea of the extremely unpleasant temperatures and pressures there. At about 25 kilometers (16 miles) above the surface, contact with the probe was lost; it had probably perished due to the harsh conditions.

Veneras 5 and 6 had very similar mission profiles and were probably also crushed during descent, but Venera 7 actually made it to the surface and became the first man-made object to return data after landing on another planet.

On December 15, 1970, the Venera 7 lander separated from its cruising stage and plunged into the planet's atmosphere. As it came down on the side of the planet that was facing the Earth, it could send its data directly to a Russian deep-space control station in Crimea.

As it descended, the scientists on the ground discovered that, due to a malfunctioning mechanical switch, the probe was only transmitting a single channel of data containing temperature readings. This was a disappointment, although they were later able to derive the atmospheric pressures from the temperature measurements.

Venera 7 continued to transmit temperature data down to an altitude of about 10 meters (33 feet), when another disaster struck. Suddenly the probe's parachute somehow detached or broke, and the spacecraft crashed on to the surface of Venus.

The mission seemed to be over, as the ground station was receiving nothing but background noise from the emptiness of space. Nevertheless, the station kept on recording, in the hope of eventually being able to decipher a possible signal from Venera 7 out of the radio noise.

To their amazement, they indeed managed to filter out a very weak signal that had come from the probe on Venus! For about 23 minutes after hitting the surface rather hard, Venera 7 had continued to transmit temperature data. The signal had been very weak, because the probe had fallen with its antennas pointed away from Earth.

The tough lander had managed to sent us the first in-situ measurements from the surface of another planet. The data proved that the surface was very hot, something that had already been speculated; Venera recorded temperatures between 237 and 246 degrees Celsius (460 and 475 degrees Fahrenheit), high enough to melt lead and zinc. The atmospheric pressure at the landing site was calculated to be around 93 times higher than on the surface of the Earth.

Venera 8 more or less repeated the feat of Venera 7, and transmitted data from the surface for 50 minutes. It confirmed the high Venus surface temperature and pressure measured by its predecessor. Like Venera 7, it did not return any photographs of the Venus surface, but Venera 8 managed to measure the amount of light that penetrated the thick atmosphere. It turned out to be similar to the amount of light on Earth on an overcast day.

FIGURE 7.4 A thick atmosphere with lots of clouds obscures the surface Venus, as shown in this image by Mariner 10. [NASA]

On February 5, 1974, NASA's Mariner 10 passed Venus on its way to Mercury. It zoomed by at a closest range of 5,768 kilometer (3,585 miles) and returned the first close-up images of Venus. This also marked the first time a spacecraft used a gravity assist swing-by from one planet to help it reach another.

In October 1975 the Venera 9 lander reached Venus and was separated from its orbiter. This lander had been specifically adapted to the now known harsh conditions. It had a cooling system with circulating fluid that enabled it to survive for 53 minutes after landing. Most of the instrumentation and electronics were contained in a hermetically sealed pressure vessel for maximum protection.

During descent, three parachutes and a disk-shaped structure acting as an aerodynamic drag brake ensured a soft landing, further cushioned by a compressible metal structure on the bottom.

Venera 9 was the first to show us what the surface of Venus looked like. Successful television photography revealed dust-free air and a variety of rocks of 30 to 40 centimeters (12 to 16 inches) in diameter. The same year, Venera 10 showed large pancake-shaped and weathered rocks that looked like slabs of lava. Did this mean that some volcanoes were still active on Venus?

In 1978 NASA sent two Pioneer probes to Venus, one to orbit the planet and the other to deliver four smaller probes. NASA's Pioneer 12, also known as Pioneer–Venus 1, reached the planet in December 1978. Its highly elliptical orbit around Venus brought it to within 160 kilometers (100 miles) of the surface to do radar mapping, study the clouds and make readings of the magnetosphere.

The radar system on board Pioneer 12 was able to look through the murky atmosphere. The maps that were sent back showed impressive plateaus, towering volcanoes and broad valleys larger than similar features on Earth. In the north of the planet the spacecraft spotted a mountain chain larger than the Himalayas. There were also some very large craters, that may have been gigantic volcanoes, but unfortunately the images were not detailed enough to find out.

Pioneer 13, also called Pioneer–Venus 2, carried four small probes that were dropped into the atmosphere of Venus in early December 1978. To survive the free-fall through the dense, corrosive atmosphere, the scientific instruments on board the probes had to be encased in titanium spheres. Casting and machining the titanium to the critical tolerances required more than 9 months for each container. "The project was like sculpting in titanium," recalled one engineer who was involved in the project.

The Pioneer 13 mothership launched one large 315-kilogram (694-

pound) probe toward the surface of the planet in November 1978. Instruments peered through a small window, made from diamond to be able to withstand the extreme outside pressures. A couple of days later three smaller 75-kilogram (165-pound) probes followed. The entry probes dove into the atmosphere and gathered data at different locations until they crashed onto the surface.

One of the small probes actually survived the hard landing and transmitted data for 68 minutes from the surface. The Pioneer–Venus 2 mothership also arrived at Venus in December, entering the upper atmosphere as a probe and burning up as it went down.

In December 1978, Veneras 11 and 12 descended through the atmosphere and reached the surface only 800 kilometers apart from each other. In addition to the same sensors as their predecessors, they also carried instruments capable of measuring lightning. Venera 11 counted an astonishing average of 25 flashes per second! Both probes reached the surface and transmitted data from there. Unfortunately, neither sent back any images because their television view ports failed to open.

Venera 13 and Venera 14 were identical spacecraft launched only 5 days apart in 1981. The spacecraft descent crafts and landers carried instruments to measure the composition of the atmosphere, monitor the spectrum of scattered sunlight, and record lightning during their descent.

Once on the surface, the twin spacecraft made panoramic pictures with their camera systems, and their mechanical arms began to drill out samples. The pieces of rock they obtained were deposited into a hermetically sealed chamber, where an X-ray fluorescence spectrometer measured their composition. Although Venera 13 landed in a highland area and Venera 14 in a lowland region, at both locations the surface was

FIGURE 7.5 The surface of Venus as seen by Venera 13.

found to be composed of alkali basalt. The surface of Venus thus appeared to be very uniform, and this was later confirmed by data from other missions.

Both landers also carried a spring-loaded arm to hit the surface and measure its compressibility. However, Venera 14 proved that Murphy's Law also works on Venus, because its arm landed on one of the ejected lens caps that had protected a camera lens during the descent. Instead of measuring the properties of the Venusian soil, it sent back data on the compressibility of the lens cap!

Veneras 15 and 16 were twin orbiters that mapped the surface of Venus with radar systems that were much better than those of Pioneer 12. The Veneras found that Venus was indeed covered with volcanic scars, but they found no active volcanoes. The data was not only studied by Soviet scientists, but also by planetologists in the USA and Europe as, for the first time, Russia shared data from its space explorations with the West.

Russia continued its exploration of Venus in 1984 with an exciting twin mission. The two spacecraft, Vegas 1 and 2, each had a mass of no less than 2,500 kilograms (5,500 pounds). In June 1985 both delivered entry probes at Venus.

The Venus entry package of each Vega craft consisted of a 2.40-meter (7.9-foot) diameter sphere identical to those of Veneras 9 through 14. Like these predecessors, the Vega landers were to study the atmosphere as well as the surface.

In addition to temperature and pressure measuring instruments, the descent probes carried an ultraviolet spectrometer for measurement of atmospheric constituents, an instrument to measure the concentration of water vapor, and a series of instruments for studying the chemical composition of grains and drops floating in the atmosphere. After landing, small surface samples near the probes were analyzed by gamma spectroscopy and X-ray fluorescence.

Apart from the landers, each Vega entry package also included a 3.4-meter (11-foot) diameter instrumented balloon that was deployed immediately after entry into the atmosphere at an altitude of 54 kilometers (34 miles). The balloon supported a probe with a total mass of 25 kilograms (55 pounds).

A 5-kilogram (11-pound) payload of scientific instruments hung suspended on a cable 12 meters (39 feet) below the balloon. The balloons floated at an altitude of approximately 50 kilometers (31 miles), in the middle of the most active layer of the Venus cloud system. Onboard instruments measured temperature, pressure, vertical wind velocity, and visibility to indicate how many small drops of liquid were floating around in the atmosphere.

After two days, and traveling some 9,000 kilometers (5,600 miles) through the atmosphere, the Vega balloons left the night side of Venus and entered the day side. Because of the intense heat of the Sun, the balloons quickly warmed up, expanded and burst.

However, that was not the end of the Vega missions. While the balloons and entry probes were doing their work, the Vega motherships had swung by Venus and were on their way to intercept the famous Halley comet.

The first spacecraft encountered comet Halley on March 6, 1986, and the second three days later. Both flew by with a velocity of no less than 78 kilometers (48 miles) per second, Vega 1 at an estimated distance of 10,000 kilometers (6,200 miles) and Vega 2 at about 3,000 kilometers (1,900 miles). The data they gathered during their short visits about the comet's orbit was very useful for targeting the ESA Giotto comet probe, which would pass Halley even closer.

Magellan

The most revealing Venus mission yet, NASA's Magellan, left Earth in 1989 and entered orbit around Venus on August 10, 1990. It was born out of another project for a very sophisticated robotic Venus explorer, the Venus Orbiting Imaging Radar, VOIR, that was cancelled because it was found to be too expensive. However, smart engineers combined hardware that had already been built for this orbiter with spare parts from other space explorers such as Voyager, Ulysses and Viking to create a new spacecraft that was named Magellan. Even equipment that had already been put in the National Air and Space Museum in Washington was incorporated to keep the costs down. Because the spacecraft also somewhat resembled a flower, it was nick-named "Secondhand Rose."

Magellan was the first interplanetary spacecraft to be launched by the Space Shuttle. Once out of the Orbiter's cargo bay, a solid propellant rocket motor called the Inertial Upper Stage (IUS) pushed Magellan out of Earth orbit. It looped around the Sun one and a half times before arriving at Venus 15 months after launch. Another solid propellant motor then fired to place the spacecraft in orbit around the planet.

Its initial orbit over the planet's poles was highly elliptical, taking Magellan as close as 294 kilometers (182 miles) to the surface and as far away as 8,500 kilometers (5,300 miles). Later, once the main part of its mission was over, Magellan was placed in a much lower orbit of 180 by 541 kilometers (112 by 336 miles). This was done by using the then still

FIGURE 7.6 The Magellan spacecraft is prepared for launch at the Kennedy Space Center. [NASA]

experimental and untried technique of aerobraking: carefully planned maneuvers sent Magellan dipping into the atmosphere at the lowest part of each orbit. The atmospheric drag slowed the spacecraft down, diminishing its average orbital altitude and also making the orbit more circular.

Magellan carried a very powerful radar to penetrate the opaque atmosphere and obtain spectacular images of Earth's cloud-shrouded sister planet. During the part of its initial orbit closest to Venus, the radar mapper imaged a swath of the planet's surface 17 to 28 kilometers (10 to 17 miles) wide and 1,600 kilometers (1,000 miles) long. At the orbit's furthest point, the spacecraft radioed the long image strip captured during that orbit, nicknamed a "noodle," back to Earth. The smart and economical design of the spacecraft allowed the use of the same large dish antenna both for the radar instrument and for sending the gathered data to Earth.

As the planet rotated under the spacecraft, Magellan collected and dispatched strip after strip of radar image data. Magellan mapped Venus more complete and in more detail than ever before. Features as small as 120 meters (400 feet) could still be discerned.

What scientists saw on them strongly reminded them of Earth; there were Californian earthquake faults, Hawaiian volcanoes, and rift valleys resembling those of East Africa and Europe's Rhine Valley. Long, parallel valleys and ridges like those of the Basin and Range province in California and Nevada were discovered. Furthermore there were impact craters, jagged quake faults and expansive lava flows similar to those on Hawaii and in the Snake River plains of Idaho.

Magellan found no less than 167 volcanoes larger than 100 kilometers (62 miles) in diameter; on Earth only the main island of Hawaii has such enormous dimensions! Furthermore, there were more than 50,000 smaller volcanoes all over the planet. Clearly volcanism had dominated the shaping of the planet's surface; however, no active volcanoes were spotted. Nor was there any trace of plate tectonics, the most important geological process on Earth. Nevertheless the small number of meteor impact craters on Venus, only about 900, implies that the surface is relatively young and thus must have been renewed recently. The fact that these craters are evenly scattered over the planet also suggests that the whole surface of Venus is about the same age!

The latest theory devised to explain this, suggests that the outer layer of Venus is much better than Earth in containing the heat from the liquid inside of the planet. On Earth water helps to make lava liquid and also lubricates the continental and oceanic shelves so that they can move over and under each other. Venus is so dry that the amount of heat that can escape through volcanoes or via movements of the surface is relatively low. Moreover, the high atmospheric pressure prevents some volcanoes from erupting, because it is pushing the magma back down.

The heat builds up, until it gets so hot that the whole exterior of

Venus cracks open and its outer layer sinks away in the hot magma beneath. In one short, cataclysmic move of cosmetic surgery, the whole planet is suddenly rejuvenated. The last time that happened may be no more than 200 to 600 million years ago; quite recent in terms of planetary age.

This process would explain why the surface composition measurements made by Venera 13, in a highland area, and Venera 14, in a lowland region, were so similar.

Magellan was also used to map the planet's gravity field by sending a constant radio signal back to Earth. Variations in the gravity field changed the spacecraft's orbital velocity, which could be seen in the received radio signal as a Doppler shift in frequency.

At the end of Magellan's mission, in October 1994, aerobraking was used to lower the probe's orbit until it plunged to the planet. Most of the spacecraft will have been vaporized by the fiery descent, but some parts probably made it to the surface.

Venus Express

Venus's next visitor will be ESA's Venus Express. The orbiter will study the atmosphere, and hopefully will make us better understand why Earth and Venus have developed so radically differently.

Scientists want to clarify the origin of the violent winds that continuously blow around the planet. The atmosphere goes around Venus about once in four days, while the planet itself actually takes about eight Earth months to make one rotation around its axis. The result of the atmospheric "super-rotation" is that the winds blow with hurricane velocities.

Venus Express may also finally ascertain whether there are active volcanoes on the planet. There is a bit of friendly competition going on between the various scientists involved in the project, to see who will be first to locate an active volcano. The camera that is primarily intended to image the movements of the clouds may also be able to find infrared signatures of recent lava flows. One spectrometer instrument is going to measure concentrations of sulfur compounds in the atmosphere, which may point out spots with active volcanoes. Another spectroscope instrument will measure the temperature of the surface of Venus and may find hot lava that way.

Also intriguing is a mysterious layer in the atmosphere that blocks ultraviolet radiation. We currently do not know its composition, but one theory suggests absorption by micro-organisms. High above the lethal areas of the atmosphere, some form of microscopic life may just be able to survive. After all, at an altitude of 50 kilometers (30 miles) the pressure is about the same as on the surface of the Earth and the temperature is a pleasant 20 degrees Celsius (70 degrees Fahrenheit). The measurements made by Venus Express can help us to find out.

Other important issues that Venus Express is going to investigate are: the formation and evolution of clouds and haze at different altitudes; the origin of mysterious ultraviolet marks at the clouds tops; the processes that govern the chemical state of the atmosphere; the role greenhouse effect plays in the global evolution of the Venusian climate; the processes by which gasses escape from the atmosphere; and the cause of the global volcanic resurfacing of Venus a couple of hundred million years ago.

The spacecraft will mainly study the atmosphere, but is also going to see the surface; even though the atmosphere is opaque for visible light, there is a kind of electromagnetic "window" in the infrared. This was confirmed by the Galileo spacecraft when it flew past Venus on its way to Jupiter, and Venus Express is going to be the first spacecraft to take advantage of it.

Venus Express is reusing the basic spacecraft design of ESA's Mars Express probe and carries the kind of instruments originally developed for Mars Express and Rosetta. By reusing and adapting already existing spacecraft equipment and scientific instruments, Venus Express was developed and built very quickly and for relatively little money. It took only three years to go from project approval to the launch pad.

However, when Venus Express was all set to launch, ESA engineers discovered that small particles of insulation material from the rocket fairing had fallen onto the spacecraft. The launch had to be delayed, so that the engineers could remove the spacecraft from its upper stage booster and clean the orbiter. Fortunately, the pieces of material were found to be large enough to see and remove with tweezers and vacuum cleaners.

On November 9, 2005, the Soyuz Fregat launcher that carried Venus Express finally left the launch pad. The spacecraft was scheduled to arrive at its destination in April 2006 (while this book was being produced).

MARS

Mars, with its distinctive orange-red hue and its periodic loops against the background stars (due to the combined motions of the Earth and Mars in their orbits around the Sun), has mystified and intrigued people for as long

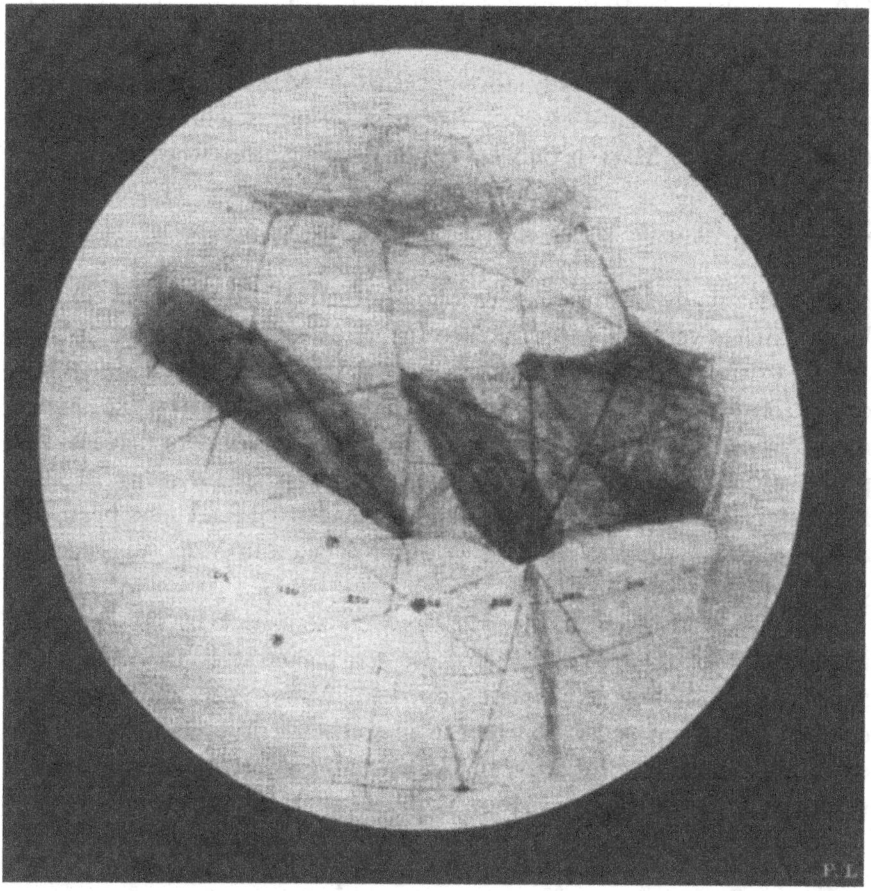

FIGURE 7.7 Percival Lowell was convinced that he could see channels on the Surface of Mars, as shown in this drawing made at his observatory. They were later proven to be optical illusions. [Lowell Observatory]

as can be remembered. The ancient Greeks, and later the Romans, linked it to the god of war.

In more modern times Mars became the subject of creative observers who believed they could see artificial channels on the surface of the planet through their telescopes. In their opinion the lines they discovered were simply too even and straight to be of natural origin, which led to the conclusion that Mars must be inhabited by intelligent beings.

For many writers this was the inspiration for stories about malevolent Martians that meant no good to planet Earth, often wanting to destroy or to invade our world. This gave Mars a rather bad image with the public, something that the innocent red planet really didn't deserve.

H.G. Wells's *The War of the Worlds* of 1898 had a particularly huge influence when, in 1938, Orson Welles adapted it as a radio play. The broadcast led to an enormous panic in the United States, as many unsuspecting listeners really thought that a Martian invasion was in progress. Partly because of such fantasies, serious scientists also turned their attention to Mars, if only to invalidate the horror stories.

Stumbling toward Mars

Mars had already become a target for both the Soviet Union and the United States very early in the space age. After some attempts that failed during or just after launch, Russia managed to send Mars 1 on its way on November 1, 1962. As with so many of those early exploration missions into the Solar System, Mars 1 was not a complete success. The probe did zoom past Mars as planned and passed within 200,000 kilometers (124,000 miles), but because radio contact had been lost three months earlier no pictures or other data about Mars was received on Earth.

The first 21 detailed images of Mars were sent back by the American Mariner 4 probe, after it flew past the red planet at 10,000 kilometers (6,200 miles) in July 1965. The hoped for Mars canals and large areas covered with alien vegetation that even some scientists had expected to exist were nowhere to be seen. In fact, the Martian surface looked surprisingly like that of the Moon with its craters and hills.

Both the general public as well as scientists were disappointed in the dull, monotonous landscape that Mariner 4 presented them with.

Mariners 6 and 7, which flew by Mars only a few days before Neil Armstrong and Edwin Aldrin stepped on the Moon, did however show a much more spectacular side of Mars. On the new images we saw extremely deep canyons, vast areas covered with sand dunes and mysterious landscapes that seemed to be sculpted by large amounts of water that one time flowed over the now dry planet. And there was clearly much more to be discovered, because the pictures of the two spacecraft together only covered about 20 percent of the surface of the planet.

In November 1971, Mariner 9 ignited its thrusters to become the first spacecraft to go in a continuous orbit around Mars. Instead of flying by and snapping only a few pictures, as its predecessors had done, it mapped the entire planet. Mariner 9 sent back no less than 7,329 images and gave us a remarkable overview of what appeared to be a very interesting planet after all.

When it arrived, Mariner 9 found the whole planet to be covered in a gigantic dust storm. Once the dust cleared, scientists were surprised to find four enormous dead volcanoes, each much wider and higher than the largest volcano on Earth. The most spectacular of them all was Olympus Mons, a giant that rises 25 kilometers (16 miles) above the surrounding landscape. It is the largest volcano in the entire Solar System.

Another spectacular surprise was Valles Marineris, a vast system of canyons that dwarfs the Grand Canyon. It is about 4,000 kilometers (2,500 miles) long and at some places no less than 7 kilometers (4 miles) deep. Valles Marineris appears to be an enormous fracture, possible the result of deformation of the surface by the four large volcanoes.

The Soviets tried again in 1971 with Mars 2 and Mars 3, both consisting of an orbiter and a lander. The Mars 2 lander hit the ground too hard, but the Mars 3 lander made the first successful landing on the Martian surface.

Unfortunately contact was lost after a mere 21 seconds, even before any picture could be sent to Earth. Later analyses showed that the unfortunate probe had probably landed in the middle of one of the gigantic dust storms that harass Mars.

The missions of the Mars 2 and Mars 3 orbiters were, however, a complete success. For months they sent a steady stream of scientific data back to the scientists on Earth.

Further missions were attempted in 1973, when Russia launched the Mars 4 and Mars 5 orbiters as well as the Mars 6 lander. Again they ran out of luck, as Mars 4 passed the planet instead of going into orbit around it and contact with Mars 6 was lost just before landing.

Mars 5 did get into orbit around Mars, but only operated for a short time due to an error in a computer chip. Furthermore, the few images it managed to return were of a much lower quality than those already sent by Mariner 9.

Viking invasion

The most spectacular missions to Mars until then arrived in 1976, when the USA put its two Viking spacecraft in orbit around the planet. As Mars was at that time covered by an enormous dust storm, the landers of the twin spacecraft had to remain docked to their motherships longer than planned. However, as soon as the storm had subsided, the two landers went down, Viking 1 on July 20 and Viking 2 on September 3, and made successful landings on the dusty red surface. The Viking landers sent back the first pictures of the surface of Mars.

The images showed rusty brown plains covered with dust and rocks of

FIGURE 7.8 The first picture made from the surface of Mars by Viking 1 in 1976. [NASA]

all sizes under a pink sky due to the high amount of red dust in the atmosphere. The weather stations on board the landers measured that the temperatures at the landing sites varied between –85 and –3 degrees Celsius (–121 and +27 degrees Fahrenheit). A light wind was blowing, which changed direction every day according to a fixed, predictable pattern.

The atmospheric pressure on the surface was found to be only 8 millibars, very low in comparison with the average 1,000-millibar pressure we are used to on Earth at sea level. Viking lander measurements also showed that the atmosphere contains 95 percent carbon dioxide, 3 percent nitrogen and 2 percent argon, oxygen, carbon monoxide and other gasses. Very different from the Earth's atmosphere, which contains 78 percent nitrogen, 21 percent oxygen, less than 1 percent carbon dioxide and an even lower amount of other gasses. The humidity of the Mars atmosphere at the surface was found to be extremely low.

The landers also investigated the soil by scooping up some sample material and dropping it in an ingenious, automatic mini-laboratory. The ground was found to contain silicon (15 to 30 percent), iron (16 percent), calcium (3 to 8 percent), aluminum (2 to 7 percent) and a little bit of titanium. Its red color was proven to be caused by iron oxide, also known as rust.

Scientists had furthermore prepared the automatic laboratories with a range of experiments to look for (bacterial) life on Mars (the Viking mission was the first ever to implement Planetary Protection measures). Unfortunately all their results could be explained by non-biological chemical reactions, and so no evidence of anything being alive on Mars was found.

Disaster strikes

After Vikings 1 and 2, it took no less than 12 years before we returned to Mars. The Soviets then launched their two Phobos probes. The target for these missions was however not Mars itself, but Phobos, one of the two small moons orbiting Mars. The Phobos probes would both slowly orbit this moon, and Phobos 2 would drop two small landers on the surface for further investigation.

Tragically, Phobos 1 was lost before it reached its destination due to a faulty command sent by the mission control on Earth. In an entirely unexpected reaction, Phobos 1 switched itself off completely and refused to be powered up again.

Investigations later showed that the disastrous software routine that enabled the inadvertent shutoff command had originated in the testing phase back on Earth. For the test model of the Phobos probe temporary commands had been built into the software to enable test engineers to quickly shut down parts or even the entire spacecraft. For the actual flight software, programmers forgot to delete the now unnecessary and potentially dangerous software lines. Murphy's Law – which says that if something can be done in the wrong way, someone is bound to do just that – did the rest.

Phobos 2 had approached the moon Phobos to within a distance of only 10 kilometers (6 miles) when contact with this probe was also suddenly lost. This happened just after Phobos 2 had made pictures of Mars and was commanded to aim its main antenna back to Earth again. Nothing was ever heard from the spacecraft; probably control problems had claimed this unfortunate space explorer too.

NASA was also unsuccessful in returning to Mars with the Mars Observer orbiter in September 1992. Just before it arrived at its destination, contact was unexpectedly lost and never re-established. Later it was determined that the most likely cause for the loss of Mars Observer was a design error in the propulsion subsystem, which had probably led to a violent explosion of the probe when the two automatically combusting propellants got mixed inside the piping to the thrusters.

Return to the red planet

NASA's Mars Global Surveyor was launched in November 1996 and arrived at Mars about a year later. This was the first spacecraft that had been designed to use aerobraking as a fundamental part of its mission. Aerobraking had been successfully demonstrated in the final days of the Magellan mission to Venus, but had never been used previously as a so-called mission critical step.

Mars Global Surveyor was initially inserted into a highly elliptical orbit. Then it used its solar panels to aerobrake in the upper layers of the Martian atmosphere as it dipped down during the low point of its orbit. It even had special panels at the ends of its solar panels to maximize drag.

However, a problem arose when one of the two solar panels hinged itself past its designed position. It appeared that the latch on the solar panel had cracked! Fortunately the mission operators were able to design a new, lower-drag aerobraking procedure that exerted less stress on the damaged solar panel. The disadvantage was that the aerobraking procedure now took much longer than anticipated: the mapping phase was originally planned to begin in the spring of 1998, but because of the delay it did not start until over a year later, in April 1999.

However, the rest of the mission went very much according to plan. At the time of writing Mars Global Surveyor is still sending out a steady stream of very detailed pictures and contour mapping data from its laser altimeter. Its images have already led to new insights, probably the most spectacular of which are indications of the existence of underground water reservoirs that sometimes break through the walls of large craters.

Because of Mars Global Surveyor, we now know that Mars is a fascinating and very complex world – a true paradise for geologists and geophysicists, and a planet that, earlier in its existence, has been a warmer, wetter and more life-friendly place.

Russia made another attempt to reach Mars in 1997 by launching the Mars '96 mission, in which a total of no less than 20 countries participated. The ambitious mission consisted of an orbiter, two landers and two torpedo-like penetrators designed to descend at great speed and bury themselves deep into the Martian soil. However, the fourth rocket stage of the usually very reliable Proton launcher did not work and the probe presumably ended up on the bottom of the Pacific Ocean.

On July 4 of the same year, NASA was much more successful when its Pathfinder probe bounced over the Martian surface, safely tucked inside its airbag protection. Once the airbags were deflated and its solar panels, cameras and antennas deployed, the lander made marvelous stereoscopic images of its landing spot.

Then "Sojourner," the little rover it had carried inside, drove off for its now famous exploration of the area. The mobile mini-robot was equipped with an experimental arm that could be placed against rocks to investigate their composition.

Sojourner's travels and its encounters with fancifully named rocks like Yogi and Barnacle Bill could be followed live on the Internet. The NASA Pathfinder website received a record number of visits and for the first time made it possible for the general public to closely follow a robotic space exploration mission.

The next two NASA missions, Mars Climate Orbiter and Mars Polar Lander, were complete failures. As mentioned before, Mars Climate Orbiter was lost when rather than entering orbit as intended, it crashed due to the mixing of imperial and metric units in the software. Mars Polar Lander was also lost, possibly because of an inadvertent hard landing caused by the main rocket descent engine shutting off too early at much too high an altitude. The shock caused by deployment of the landing legs had probably been picked up by a touch-down shock sensor, which made the lander think it had already reached terra firma.

The two Deep Space 2 penetrators, which accompanied Mars Polar Lander on its mission to investigate the south pole of the red planet, were also never heard from. What happened with them is still completely unclear.

For the Mars Climate Orbiter and the Mars Polar Lander/Deep Space 2 mission, NASA had tried to work on budgets that were too low and time schemes that were too ambitious, which left insufficient money and time for proper design verification and testing.

NASA's Mars program fortunately returned to success with the launch of Mars Odyssey in 2001 (named after the famous space movie *2001, A Space Odyssey*). After arrival in Mars orbit it started to globally map the composition of the surface, studying the kind of minerals that are present and looking for hydrogen, which could indicate the presence of water, in the shallow subsurface. It also carried an instrument to measure the radiation in space near Mars, which would help us to find out how dangerous this could be for future astronauts traveling to the planet.

The results of Mars Odyssey, which is still working well at the time of writing, indicate that there may be water or water-ice hidden under the

FIGURE 7.9 Artist impression of NASA's Mars Odyssey spacecraft in orbit around Mars. The instrument on the long mast is the Gamma-Ray Spectrometer that is used to map soil composition. [NASA]

Martian surface. Many of the minerals it detected also indicate that lots of water once flowed over Mars. The radiation levels that Odyssey measured do not seem to be too dangerous for astronauts or too difficult to handle for crewed spacecraft designers, so one day humans should be able to go and land on Mars.

ESA's Mars Express was the next spacecraft to enter orbit. It was Europe's first ever Mars mission, and although the Beagle 2 lander it carried was never heard from again after detachment from the orbiter, the mission of Mars Express itself has become a huge success.

It has found indisputable proof for the presence of water-ice just under the Martian surface, and its stereoscopic color imager has mapped the entire surface in unprecedented detail. Mars Express has shown that the rocks found by the later NASA MER Marsrovers, which are thought to have been formed in the presence of water, can be found all over the planet. These "hydrated" minerals, so called because they contain water in

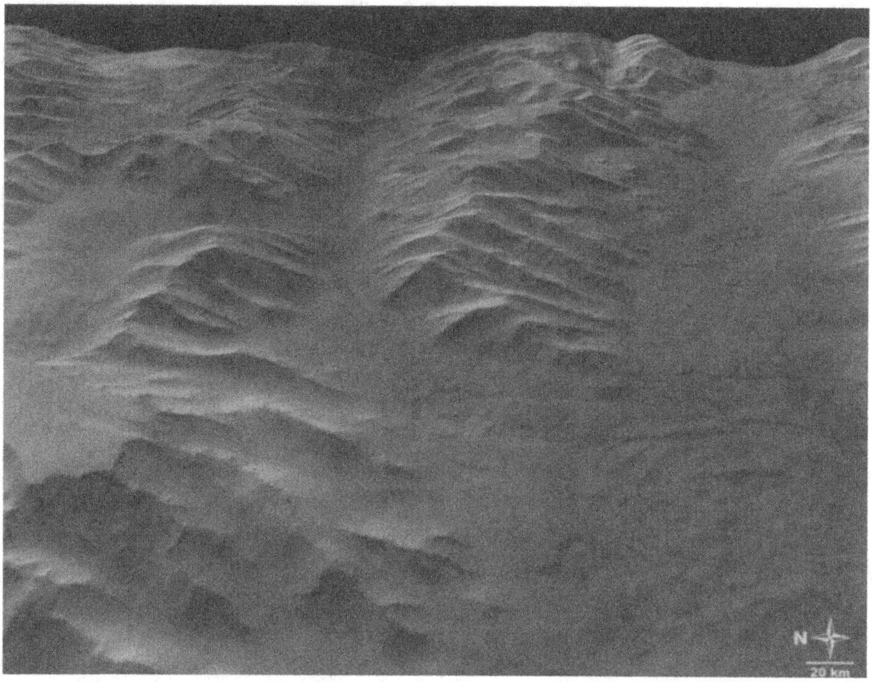

FIGURE 7.10 A 3-D view of Titonium Chasma, based on images made by the High Resolution Stereo Camera of Mars Express. [ESA]

their crystalline structure, provide a clear "mineralogical" record of water-related processes on Mars.

Next to the finding of large amounts of water-ice, the most spectacular news was probably the discovery of traces of methane (natural gas) in the Martian atmosphere. This gas should be fairly quickly lost from the planet unless it is continually replenished. It could be that the methane is being expelled from deep inside Mars, or is produced out of the atmosphere's carbon dioxide by the Sun's ultraviolet radiation. Perhaps a large comet full of frozen methane has recently crashed on the planet. However, the most tantalizing possibility is that the methane is being produced by microbes, just as on Earth.

Unfortunately none of the orbiters and landers we have sent to Mars till now carried the right instruments to find the cause for the existence of methane in the atmosphere. Future Mars orbiters and landers may take experiments that are specially designed to do that.

Shortly after the arrival of Mars Express, Christmas 2003, the famous NASA Mars Exploration Rovers, the MERs, arrived on Mars. Each of the spacecraft, named "Spirit" and "Opportunity" as the result of a naming

FIGURE 7.11 This picture of part of the Bonneville crater was made by NASA's Marsrover Spirit. [NASA]

competition for children, carried a sophisticated rover much larger than the earlier Sojourner.

As already described in Chapter 5 ("Instruments of Science"), the twin rovers have found ample evidence of the large role water has played in the history of Mars, and they continue to do so. Although they were originally designed to operate for three months, at the time of writing they have already been driving around six times as long.

Mars is currently under surveillance by three orbiters (Mars Global Surveyor, Mars Odyssey and Mars Express) and two rovers. No other planet apart from Earth has ever been studied in such detail. We now see Mars as much more than a dry, barren and rocky planet, a red and larger version of the Moon. Mars is a real world, with canyons, mountains, volcanoes, beaches, riverbeds and polar caps. It is a geological paradise where no features are obstructed by forests, water or cities.

At one time, it must have had a lot of water on its surface, probably even real oceans. There may have been life in that water. Recent studies indicate that Mars has even known ice ages, which may have covered large parts of the planet in snow and ice.

Computer renderings of wet and ice-covered ancient versions of Mars, such as made by Kees Veenenbos, remind us of just how Earth-like Mars may once have been. It is not hard to imagine ourselves walking through the canyons, chipping off pieces of rock to look for fossils. After Earth, Mars is the most hospitable place in the Solar System.

The year 2006 should see the arrival of NASA's Mars Reconnaissance Orbiter (MRO), a satellite that will be able to see objects on Mars as small as a dining table. MRO will try to find out more about the history of water on Mars. How much was there? For how long did it freely flow over the surface? Its instruments will make extreme close-up images of the Martian surface, analyze minerals, look for subsurface water, trace how much water is circulating in the atmosphere, and monitor the weather on Mars.

It will determine whether there are deposits of minerals that only form in water over long periods of time, look for shorelines of ancient seas and lakes and analyze sediment deposits placed by flowing water. It will also be able to tell if the underground ice discovered by Mars Odyssey and Mars Express is the top surface of a large ice deposit or whether it is only a shallow layer.

The MRO mission will send us several times more data about Mars than all previous missions combined, transmitting about 10 times as much data per minute as any previous Mars spacecraft.

The next Marslander is scheduled to be launched in 2007, when NASA's Phoenix mission takes off. Phoenix is targeted to touch down softly in the northern polar region, and will be the first spacecraft to do so. Rather than using airbags, a similar rocket system as was used for the Viking landers will ensure a soft and controlled landing.

Phoenix is a stationary lander, but it has a robotic arm to dig down into the soil and deliver samples to sophisticated instruments on the lander's deck. This on board mini-laboratory is specifically designed to measure volatiles, such as water and organic molecules.

Like a phoenix it will rise from the ashes of two earlier Mars exploration missions, namely the crashed 1999 Mars Polar Lander and the 2001 Mars Surveyor lander that was mothballed in 2000. Development of the Surveyor lander was stopped as a result of investigations into the failure of the Mars Polar Lander, which indicated that the Surveyor lander mission also involved risks that were too high due to insufficient testing and management overview. For Phoenix, the management and testing rules and requirements have been tightened, and the financial budget has been set at a higher, more realistic level that enables proper development procedures.

After Mars Reconnaissance Orbiter and Phoenix, it is planned to send more landers, rovers and even missions to grab some surface samples and bring them back to Earth. The USA, Europe and Russia also aim to launch crewed missions to Mars sometime after 2030. It seems that Mars will remain in the spotlight for at least another 30 years.

Mercury, a Scorched Planet

Mariner 10, the spacecraft that first flew by Venus and used that planet's gravity to sling-shot further on to Mercury, is still the only spacecraft that has ever visited this little world closest to the Sun. It made three quick flybys of Mercury in March and September 1974 and March 1975.

The orbit maneuvers needed to achieve this were very tricky to calculate, because each time Mariner 10 flew around the Sun and re-crossed Mercury's orbit, the planet had also to be at exactly that location! After the last visit, Mariner 10's propellant was depleted and the probe became uncontrollable.

Mariner 10 sent back the first close-up images of Mercury, and showed that this barren world looks remarkably similar to the Moon. Mercury appears dark gray and is covered by impact craters. One of them, the Caloris Basin, is larger than France. The shock wave of the colossal impact that created it traveled all the way around the planet and created an entire mountain chain on the other side!

Now Mercury is dead; there is no water, virtually no atmosphere and no geological activity. The many craters on the surface testify that its surface is old and for billions of years has not been geologically renewed. Earth, and perhaps Venus, still have hot, molten cores; but similar to the way that small stones in a fire cool down more quickly than large ones, so has tiny Mercury cooled down long ago.

Mercury has a diameter of only 4,880 kilometers (3,030 miles), just a bit more than a third of that of the Earth. The many ridges, some with a height of more than 3 kilometers (2 miles) probably formed when the core cooled down and shrunk. Like the skin of a dried-out plum, it made the surface of Mercury crack and wrinkle. If there was any volcanism, it would have been stopped at that time as the contracting crust closed-off the magma conduits.

As mentioned above, at the time of writing NASA's Messenger spacecraft is its way to explore Mercury from orbit for the first time. ESA's BepiColombo with its two orbiters – one European and one Japanese – should follow around 2012.

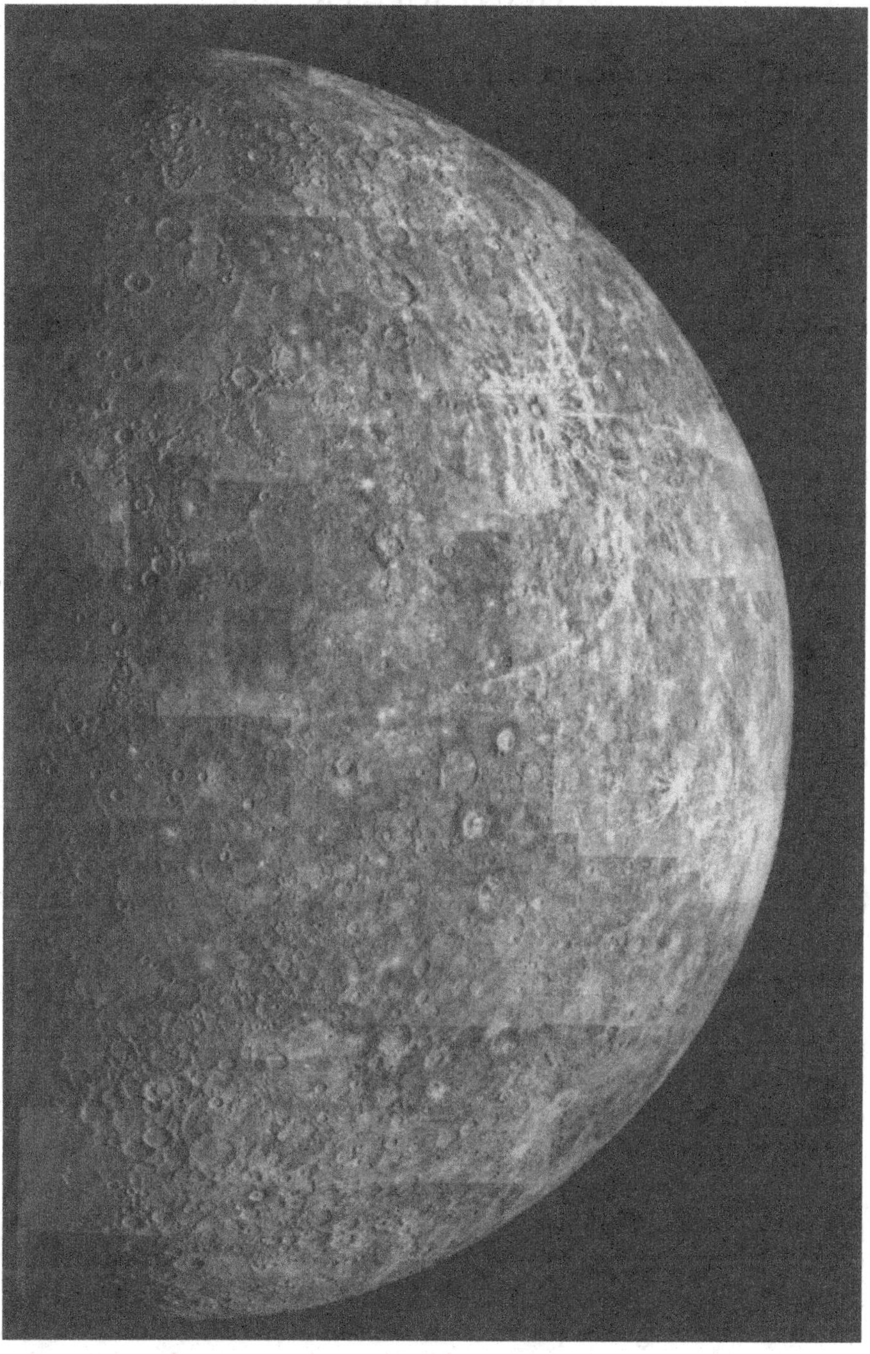

FIGURE 7.12 This mosaic image of Mercury was constructed from pictures made by Mariner 10. [NASA]

Chapter 7

GIANT JUPITER

Jupiter is the largest planet in the Solar System, with a maximum diameter of 142,800 kilometers (88,700 miles); 11 times that of the Earth. It was the first of the giant outer planets we reached when the 258-kilogram (569-pound) Pioneer 10 flew by in 1973. The probe measured the atmosphere and magnetosphere, and made over 300 images.

We quickly found out that Jupiter has a very strong magnetic field that is very good in capturing charged particles from its own moon Io and from the Sun. The radiation levels caused by these particles nearly ended the life of Pioneer 10, when during its closest approach it had to endure some 500 times the amount of radiation that would be instantaneously deadly to humans. Fortunately the robotic explorer got no nearer than 131,400 kilometers (81,700 miles), otherwise the even higher levels of radiation closer to Jupiter would surely have meant the end of it.

Another surprising feature of Jupiter found by Pioneer 10 is that the planet emits about 1.7 times as much energy as it receives from the Sun. When Jupiter was formed, the enormous pressure inside the planet heated it up, like a bicycle pump getting hot due to the compression of the air inside. This is what ignited the nuclear fusion reactions inside the Sun, but Jupiter was too small and had insufficient mass to get its core pressure high enough to become another star.

The first flickering Pioneer 10 pictures that built up on the screen of NASA scientists showed brown, yellow, orange and red cloud bands circling the planet. Then the famous "Red Spot" came into view. This oval feature had already been seen on Jupiter by the astronomer Cassini in 1665 and has been visible through telescopes ever since.

Apparently it has been rather stable for centuries. It's no wonder that astronomers had thought it to be a solid object on the surface of Jupiter, or gas swirling around an enormous mountain. However, Pioneer 10 showed that it is in fact a gigantic, perpetual storm big enough to envelope the entire Earth twice!

Jupiter consists of about 86 percent hydrogen, 14 percent helium and tiny amounts of methane, ammonia, phosphine, water, acetylene, ethane, germanium and carbon monoxide. The colored clouds consist of various compounds of ammonia and hydrogen sulfide (which, you may remember from chemistry classes at school, smells like rotten eggs). Below lie water clouds. Even though the planet could contain over 1,300 Earths, its mass is only 318 times that of our world. This indicates that Jupiter consists mostly of gasses rather than fluids and solids.

The center of the planet may consist of a small, rocky core of iron and silicates at a temperature of about 20,000 degrees Celsius (36,000 degrees Fahrenheit), surrounded by a thick shell of liquid metallic hydrogen that behaves like a metal due to the extreme pressure. At about 46,000 kilometers (27,000 miles) from the center of the planet there is probably a sudden transition from liquid metallic hydrogen to liquid molecular hydrogen (H_2, consisting of two hydrogen atoms). Above the shell of liquid molecular hydrogen is the planet's deep atmosphere.

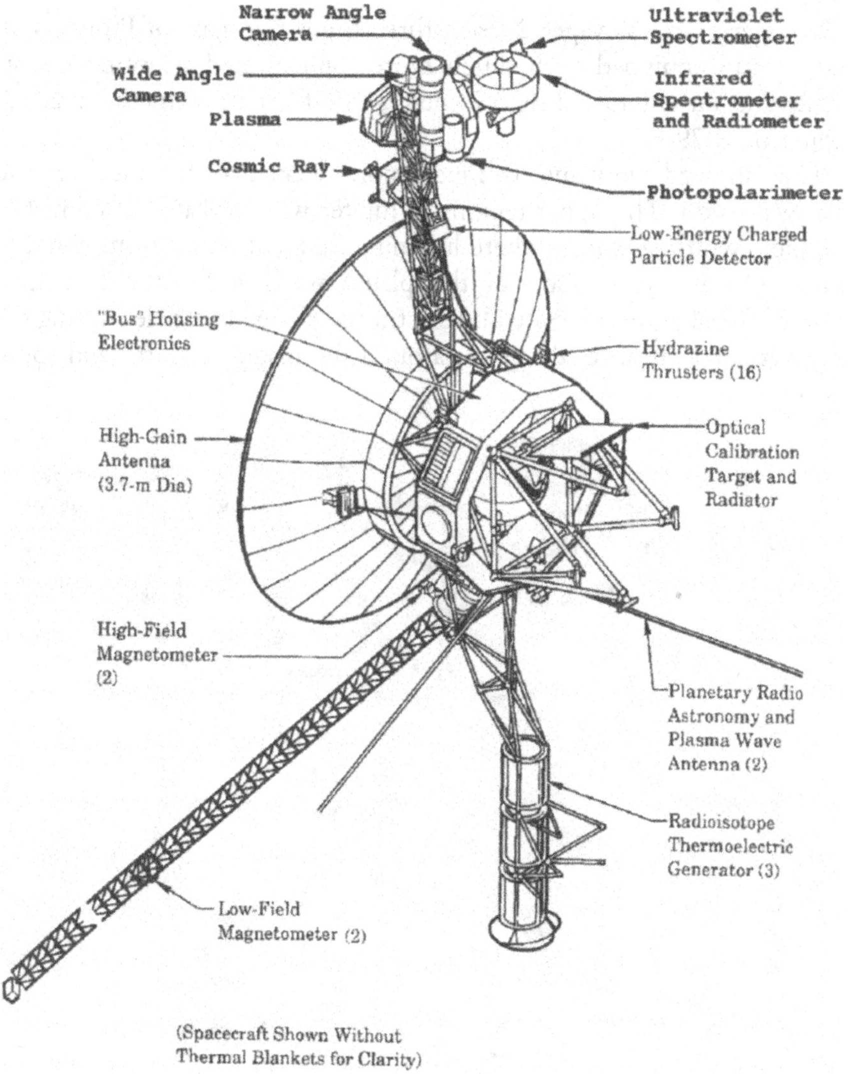

FIGURE 7.13 The Voyager spacecraft incorporated a suite of instruments to investigate the large gas planets of the Solar System. [NASA]

Pioneer 10 left Jupiter after a quick visit and is now on its way out of the Solar System. It carries a metal plaque engraved with the image of a man and a woman and information about the location of the world the spacecraft came from. At some time in the very far future other civilizations from another Solar System may intercept it. Pioneer 10 will then become our mechanical ambassador.

A year after Pioneer 10, its twin Pioneer 11 briefly visited Jupiter on its way to Saturn. It confirmed the findings of Pioneer 10 and sent us more pictures.

Voyager 1 and Voyager 2, each three times the mass of Pioneers 10 and 11 and equipped with much more sophisticated scientific instruments and an antenna dish 3.7 meters (12 feet) in diameter, reached Jupiter in 1979.

They showed the planet in much more detail than the Pioneers had managed to do. The upper regions of Jupiter were revealed to consist of complex, whirling storms where lightning frequently shot from cloud to cloud. On the night side of the planet we saw for the first time extraterrestrial auroras, created by electrically charged particles hitting the upper atmosphere in oval regions around the magnetic north and south poles.

FIGURE 7.14 The moon Io over Jupiter, as seen by Voyager 1. [NASA]

The biggest surprises came, however, from the images the Voyagers made of Jupiter's four large moons, Callisto, Ganymede, Europa and Io. Earth-based telescopes had never managed to picture them as more than small specks of light, but the Voyagers showed each to be at least as interesting as any of the planets. What also surprised scientists was that they were all completely different. Jupiter and its moons resemble a mini-solar system.

Callisto has a surface of dark, muddy ice more densely covered with craters than any other moon in the Solar System. Its surface must therefore be very old; it takes a long time to collect so many impacts. Callisto has probably remained virtually unchanged since the very birth of the Solar System.

The moon Ganymede is larger than our own, and even larger than the planet Mercury. It has mountains as tall and as wide as those on Earth.

Io is covered with layers of white, red, yellow and black sulfuric material, making the surface somewhat resemble a four-cheese pizza. Just above the rim of a picture of Io an investigator spotted a mysterious fountain-like object, which was soon determined to be the eruption plume of the first active volcano ever found beyond Earth.

As Io circles Jupiter in its orbit, it also feels the gravitational influence of the moons Europa and, to a lesser extent, Ganymede. The constantly changing pulls of the giant planet and the two other moons result in tidal forces that knead and therefore heat Io up internally. (The same happens if you knead a ball of dough; the warmth is caused by the friction between the particles inside.)

Jupiter's tidal forces heat up sulfur-rich rocks deep inside Io, releasing sulfurous gasses. When these break out to the surface, volcanoes spew them far into space. The released particles even escape the relatively low gravity of Io, go into orbit and get captured and charged by Jupiter's magnetic field. They are the main cause of the high levels of radiation around Jupiter.

Icy Europa, for the first time photographed in detail by Voyager 2, is perhaps the most intriguing of Jupiter's moons. Slightly smaller than our Moon, it is covered with a smooth icecap that completely envelops it. When the first images of Europa were received, scientists saw so few craters that they were afraid the cameras were unfocused.

We now think that the ice is several kilometers thick. Just as the interior of Io, the inside of Europa must be heated by tidal flexing, and is therefore probably liquid. That means that Europa may harbor a giant water ocean, which might even contain some form of life. Callisto and Ganymede are now thought to possibly also contain water oceans, but Europa remains the most promising. Together with Mars, this moon is therefore on the top of many planetary scientists' wish list for further investigation.

On its way out, Voyager 2 looked back toward the Sun and saw a back-lighted ring. Although very thin, it showed that Saturn is not the only planet with rings. In fact, weak rings were later found to circle Uranus and Neptune too.

Unable to brake and go into orbit, the Pioneers and Voyagers made only short flyby visits to Jupiter. In 1995 however, the Galileo orbiter arrived for a nominal four-year mission of exploration. Closely sweeping by the various moons and using their gravity to change its orbit continuously, Galileo observed the planet and the moons in unprecedented detail.

Prior to entering orbit around Jupiter, Galileo released a 339-kilogram (747-pound) probe that was to descend into Jupiter's atmosphere. Hitting the outer layers of the gas giant at an astounding 47.8 kilometers (29.7 miles) per second, the spacecraft had to endure a deceleration force 228 times that of the gravity on Earth. At the same time the temperature of its heatshield shot up to temperatures three times that of the surface of the Sun! It was like exposing the probe to the blast of a nuclear explosion at close range.

On Earth, the deceleration had been tested by putting individual components into small centrifuges. These could rotate so fast that the equipment experienced forces as high as 350 *g*, i.e. 350 times their normal weight on Earth. It's like swinging a stone attached to the end of a piece of string; the faster you rotate it, the heavier the stone will feel. The complete Galileo probe needed to be put on the much larger and more powerful centrifuge of Sandia National Laboratories. This facility resembles a thick, 24-meter-long (79-foot-long) beam with its rotational axis in the middle, the test object attached at one end and a counterweight fitted at the other side for balance. As powerful as this centrifuge is, it managed to rotate the heavy probe up to "only" 200 *g*.

To test the high-velocity flight through the hydrogen and helium atmosphere of Jupiter, NASA even had to build a special wind tunnel, the Giant Planet Facility (so named because it was designed for simulating the entries in atmospheres of giant planets such as Jupiter). This facility was used to heat and pressurize the gas inside to such levels that when it shot out in the tunnel, it resembled the gas that the Galileo probe would encounter when diving into the atmosphere. To do this, the facility consumed no less than 165,000 kilowatts of electricity, equivalent to the power needed to propel the ocean liner SS *United States* through the water at 35 knots (65 kilometers or 40 miles per hour).

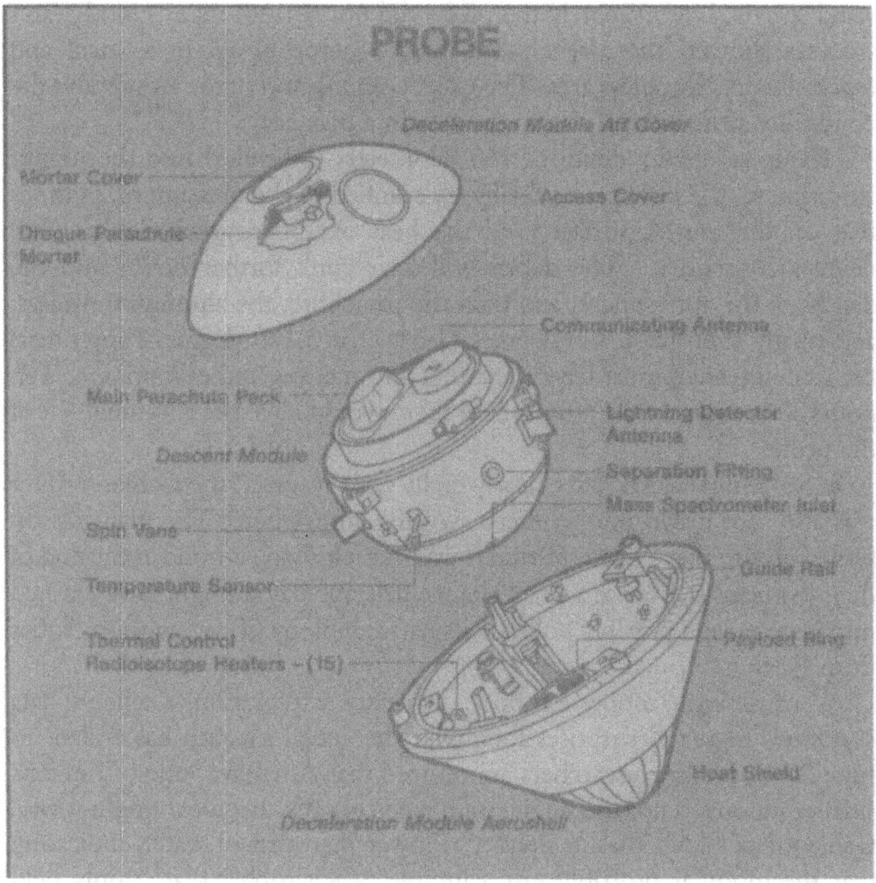

FIGURE 7.15 The Galileo atmospheric probe that went down into the swirling atmosphere of Jupiter. [NASA]

Thanks to the good design and extensive testing, the probe survived the severe heat and deceleration, ejected its protective shields and continued to descend at a speed of about 100 kilometers (62 miles) per hour.

However, later analyses of the sensor data sent back by the probe showed that its heatshield had almost burned through at the edges. Apparently the design calculations and the results of the tests performed during the development were not correct. This is not very surprising, as not much is known about what happens when a real, full-size heatshield hits an atmosphere consisting of mostly hydrogen and helium at such tremendous speeds.

Fortunately the small probe survived, thanks to the large margins taken into account in the design of its heatshield. For some 60 minutes it relayed data on the composition of Jupiter, as well as temperatures, pressures and wind speeds. Unfortunately the probe did not register many clouds; later analyses showed that, by chance, it had come down in a small and especially dry, cloudless area. The exact composition of the clouds and the conditions near them thus mostly remain a mystery.

Having reached a depth of 130 kilometers (80 miles) into the Jupiter atmosphere, the little probe finally succumbed to the pressure of 25 times that on the Earth's surface, and the heat of 150 degrees Celsius (300 degrees Fahrenheit). The debris will have sunk further down into the depths of the atmosphere; and then the parachute, the aluminum fittings, and finally even (at 1,700 degrees Celsius, or 3,100 degrees Fahrenheit) the titanium shell must have succumbed to melting and evaporation. Ten hours after entering the atmosphere there would have been nothing left of the probe.

In September 2003, after nearly eight years in orbit, the Galileo orbiter was also sent down into Jupiter's swirling atmosphere. The main reason was to ensure that a dead spacecraft was not left flying around at the end of the mission as there was a small chance that one day it might have crashed into Europa, and possibly polluted its native biology (if any) with microbes from Earth.

As usual with modern robotic planetary exploration, Galileo's data continues to reveal surprises long after the actual mission has ended. In June 2005, NASA researchers announced that Amalthea, one of Jupiter's smaller moons, is not the single, solid piece of rock that we thought it was. Galileo data shows that its density is lower than that of water, indicating that the moon is probably just a jumble of icy rubble that is only held together by its own weak gravity.

Clearly, more discoveries await at Jupiter, and we should soon return to this fascinating planet with its mysterious moons.

SATURN

The planet Saturn is best known for its magnificent rings, which are made up of millions of individual bands primarily composed of water-ice particles. The first spacecraft to see this world from close up was Pioneer 11 in 1979, which had used a gravity assist at its earlier target Jupiter to continue to the ringed planet. Only a bit more than a year later Voyager 1 arrived, and less than a year after that Voyager 2 passed by.

The Voyager images, in particular, captured the true majesty of Saturn and its rings. Rather than a small number of very broad rings, the Voyagers showed that the planet is surrounded by thousands of narrow ringlets, kept apart and prevented from dispersion by the gravitational influence of small "shepherd" moons orbiting between them.

Rather than the dull atmosphere seen through telescopes on Earth, we discovered that the complexity of Saturn's outside rivals that of Jupiter. Saturn is smaller than Jupiter, but its interior is probably very similar.

Also similar to Jupiter, the planet is surrounded by a collection of large moons, each of which is unique. Titan, the only moon in the Solar System with a real atmosphere, is the most interesting. The Voyagers discovered that Titan's atmosphere mainly consists of nitrogen and some helium and methane, and that it is covered with orange clouds made up of complex organic molecules.

Titan is a kind of early Earth in deep freeze, situated in a part of the Solar System where it is too cold for its simple organic molecules to combine and form the building blocks of living organisms. When, in a few billion years, the Sun becomes old and expands into a bloated red giant star, Titan may awake from its slumber and become just warm enough for life to evolve. However, there will be too little time for much to happen before the Sun puffs off its outer layers and the core collapses to form a white dwarf. Eventually the smaller Sun will cool down and become a dead, black star. The Earth will be long dead by then, scorched in the heat of the red giant Sun. (An intriguing story in relation to this is Stephen Baxter's novel *Titan.*)

Iapetus is a moon of which one half is snowy white while the other half is black. The fact that Iapetus always turns the same half toward its mother planet, just like our Moon, must have something to do with its strange appearance; it is the leading hemisphere in its orbit that is as black as a blackboard, while the trailing hemisphere is bright white.

Enceladus looks a bit like the Jupiter moon Europa, but only part of it is as smooth; the rest is covered in craters. It appears that the smooth area of this 500-kilometer (310-mile) diameter moon has melted relatively recently, giving the little world a partial facelift. Enceladus is Saturn's brightest and whitest satellite, and has the most reflective surface in the entire Solar System.

Mimas has a crater one-third the diameter of the moon itself, making it look like the infamous Death Star of the *Star Wars* movies ("that's no moon . . ."). The impact that caused it must have nearly destroyed Mimas.

Tethys is also scarred by an enormous crater, and just like Mimas it also consists mostly of ice.

Janus and Epimetheus share the same orbit and at one time formed a single moon together. Probably an impact, similar to those that nearly meant the end of Mimas and Tethys, broke the original satellite in two.

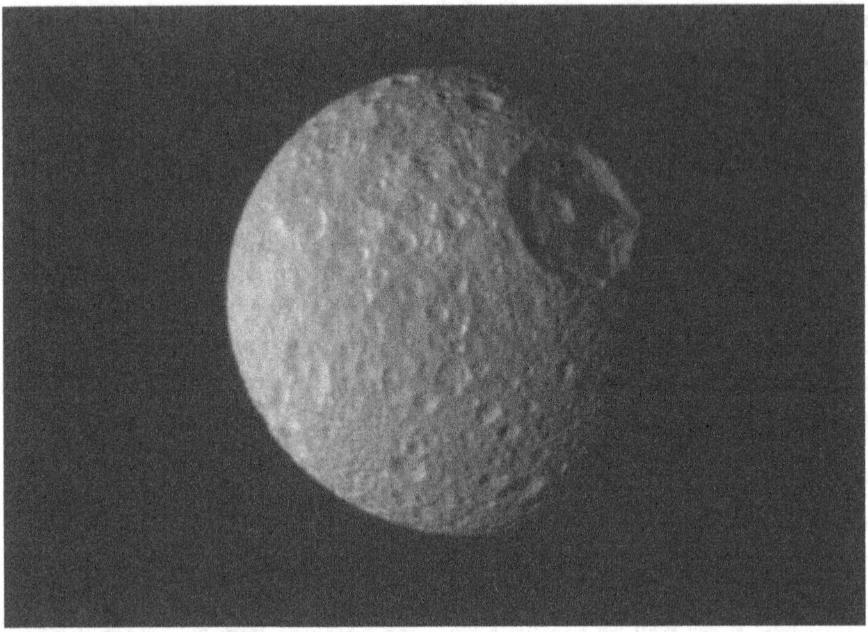

FIGURE 7.16 In this picture made by Cassini, the large crater on Mimas makes it look like the infamous Death Star of Star Wars. [NASA]

Dione, Rhea, Hyperion and Phoebe complete the collection of relatively large moons, but, just as at Jupiter, the Voyagers found many other small, often irregularly shaped bodies as well.

NASA's Cassini, the largest ever interplanetary robotic spacecraft except for the Phobos 1 and 2 probes, is currently in orbit around Saturn. Just like Galileo at Jupiter, it is not just swiftly passing by as the Voyagers did, but is on a long cruise through the Saturn system. Using the gravity of Titan, it is continuously changing its orbit, so that it can visit different moons, investigate different regions of the magnetosphere and watch Saturn and its rings from various angles.

Given that it took Cassini nearly seven years to reach its target, it would have been a waste had it only been able to take a few measurements at Saturn. Cassini was therefore equipped with an array of sophisticated instruments and cameras. It has already shown us new, breathtaking, views of the various moons and the complex ring system (see Figure 7.18).

On double-faced Iapetus it has found a long, narrow and straight ridge roughly aligned with the moon's equator crossing its black hemisphere. This has led to a new theory that, in the past, Iapetus hit one of Saturn's rings edge-on. The ridge would be the scar left by this collision, while the ring material would have spread out over half the moon and covered it

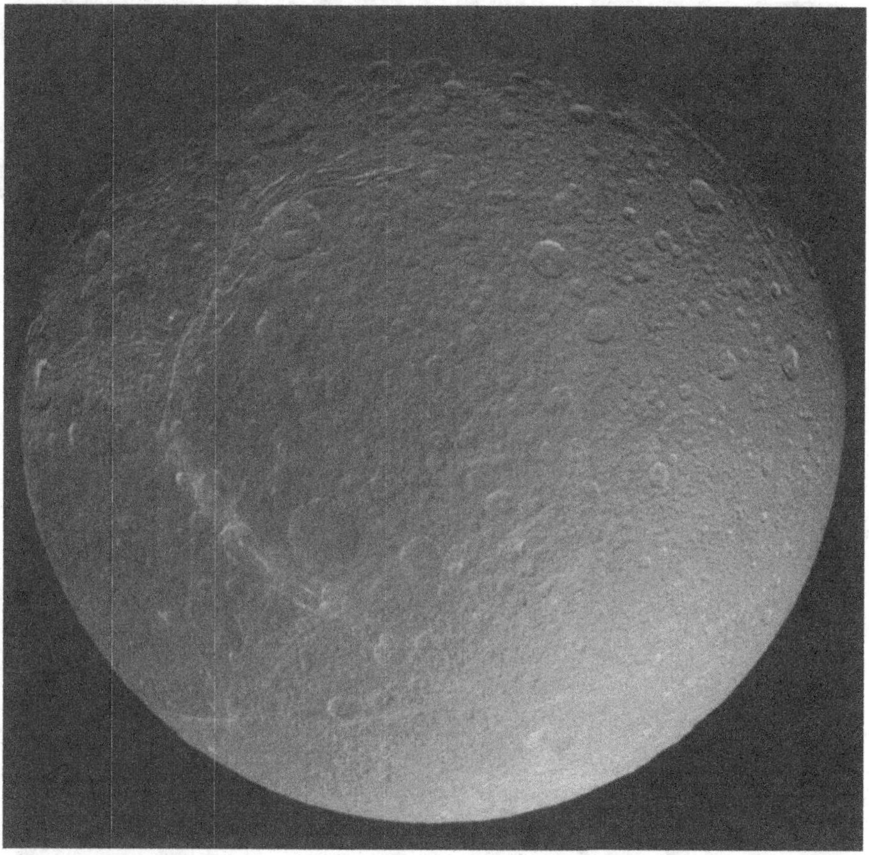

FIGURE 7.17 The surface of Dione shows bright patterns that are probably formed by material thrown out of the impact craters. This picture was made by Cassini. [NASA]

with the black material we can still see. The side of the little moon turned away from the rings kept clean, explaining the strange difference in color between the two halves of Iapetus.

When Cassini flew by Enceladus, its magnetometer measured a striking bending and oscillation of the Saturn magnetosphere, caused by ionized water vapor; Enceladus has a very thin atmosphere! Because its gravity is very low, the small moon is not able to hold on to the gasses for very long. Apparently the wispy atmosphere is continually replenished by Enceladus itself. The source for this was also discovered by Cassini: cryo-volcanism (ice volcanism)!

In addition, the Huygens probe of ESA released by Cassini at Titan in December 2004 has given us a much better idea of the composition of the moon's atmosphere. The probe found the atmosphere to be very turbulent and may even have registered lightning. It also showed that the thick

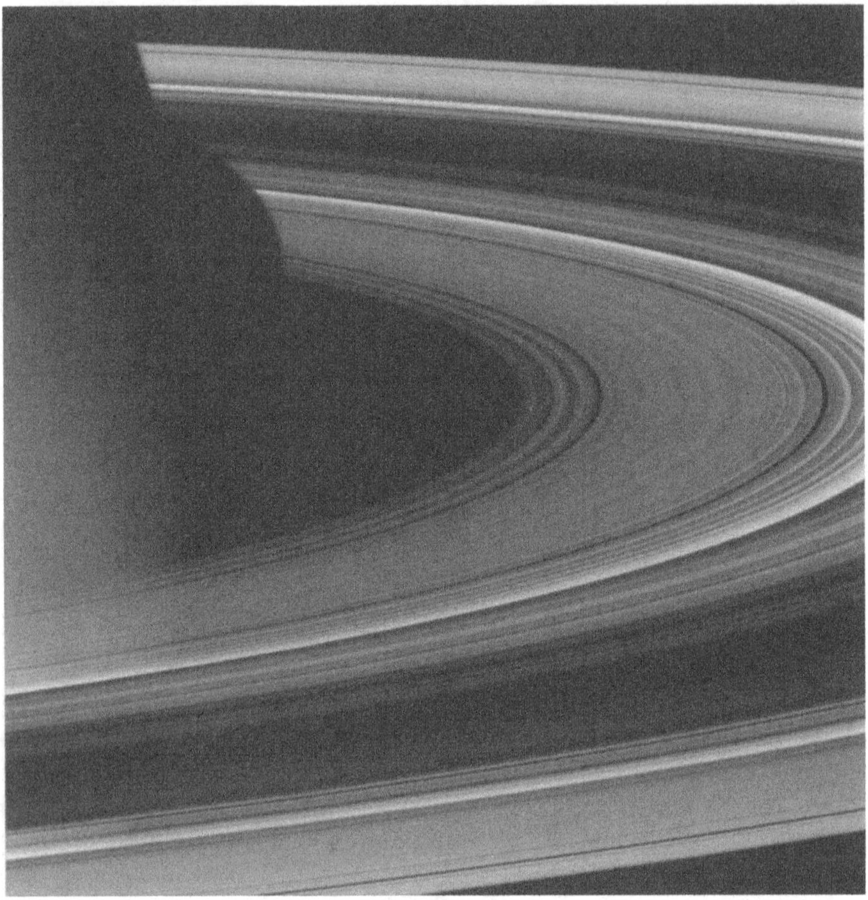

FIGURE 7.18 The rings of Saturn consist of thousands of narrow rings of dust and rocky debris, as shown in this Cassini image. [NASA]

orange haze does not disappear at an altitude of 60 kilometers (37 miles), as theorized before, but extends all the way to the surface. Fortunately it was transparent enough so that Huygens could for the first time give us a glimpse of the mysterious icy surface (see "Diving Through an Atmosphere," page 140).

Cassini and its Huygens probe discovered that Titan's surface is covered in water-ice, which methane rain washes clean from a dark goo of organic substances that collects at the bottom of hills and mountains. Because of the thicker atmosphere and the lower gravity, raindrops on Titan are probably about twice the size of those on Earth, and fall more than six times slower from the sky.

Similar processes of flow and erosion by rain, as happen on Earth, seem to have shaped the surface of Titan, but instead of being done by water, it has been done by liquid methane!

Recent infrared pictures made by Cassini through gaps in the cloudy veil over Titan's surface suggest that the methane is ejected by cryovolcanism. The lava flows probably consist mostly of water mixed with ammonia. With a much lower freezing point than pure water, the mixture is rising to the surface from within the moon, oozing out as a viscid fluid.

THE OUTER GIANTS: URANUS AND NEPTUNE

Voyager 1's planetary explorations ended at Saturn, and it is now on its way out of the Solar System. However, Voyager 2 was directed onward. Using the gravitational fields of each of the planets it visited, it traveled from Jupiter to Saturn and on to Uranus and Neptune. The alignment of all these planets had to be just right to make this game of interplanetary billiards possible – a condition that is very rare.

In 1986, just over four years after its Saturn flyby, Voyager 2 arrived at Uranus. This giant gas planet is so far away that signals from the probe, even while traveling at the speed of light, took 2 hours 45 minutes to reach us. When the images came in, scientists were somewhat disappointed with the view. They had witnessed intriguing patterns of clouds and storms on Jupiter and Saturn, but Uranus appeared to be a dull blue–green, featureless world.

The planet was also found to be much smaller than its fellow gas giants. It was only a third of the diameter of Jupiter.

The magnetic field of Uranus, as measured by Voyager 2, is much weaker than that of Jupiter or Saturn. The remarkable thing about it is, however, that this field is tilted 60 degrees with respect to the planet's rotational axis. Unlike what we are used to on Earth, a compass on Uranus would thus not indicate the geographical north!

The moons again provided a rich source of data for scientific debates, models and theories. Each of the five main moons, Miranda, Ariël, Umbriël, Titania and Oberon, show the scars of past impacts. The surfaces of all except Umbriël appear to have been severely changed since the creation of the moons; liquid water seems to have broken through their surfaces and created spectacular ice formations.

Miranda, with a diameter of only 480 kilometers (300 miles) has a vertical wall twice the height of Mount Everest! This and other extremely exotic surface features have led to speculation that this moon has once

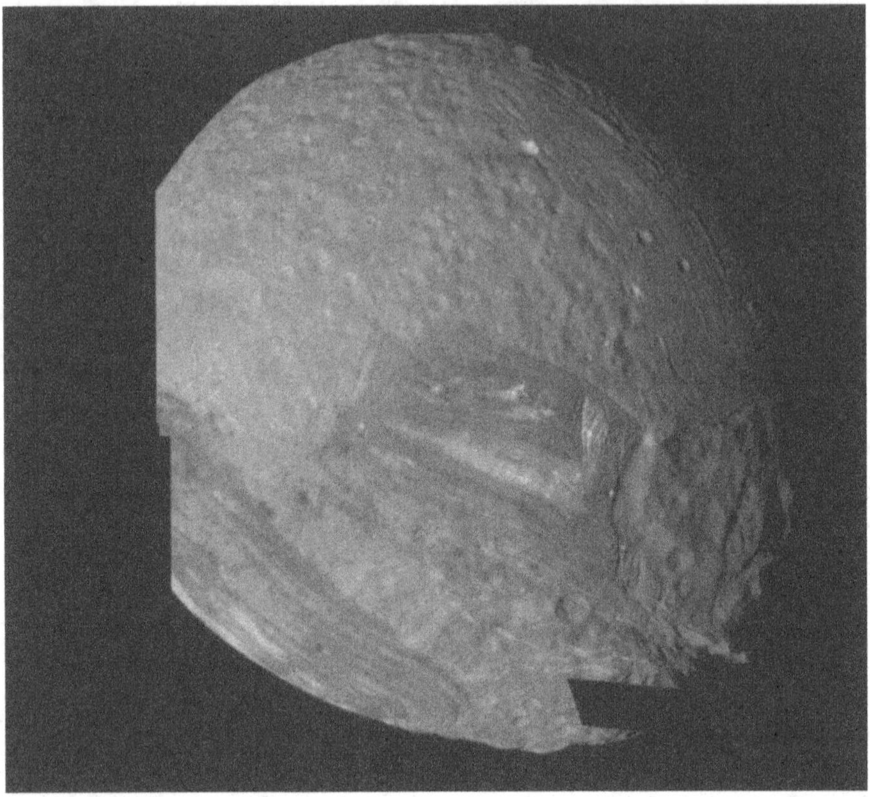

FIGURE 7.19 This mosaic of Voyager 2 images of Miranda shows the geological diversity on the surface of this 500-kilometer (300-mile) diameter moon. [NASA]

been shattered by a violent impact, then reformed out of the remaining fragments. It looks a bit like Frankenstein's monster (see Figure 7.19)!

Voyager 2 also showed that Uranus is circled by a system of thin rings, just like Jupiter. Later, similar rings would also be found at Neptune.

Voyager 2's "Grand Tour" ended with a visit to the last large planet in the Solar System, Neptune, in 1989. This gas planet is as blue and about as big as Uranus, but to the delight of the scientists it did show some clouds. The winds that propel these white wisps around the planet were found to be much faster than those at Saturn, the previous record holder for fastest storms in the Solar System.

The big moon Triton was found to be a real oddball. From Earth, astronomers had already discovered that Triton rotates around Neptune from east to west, while nearly all other moons in the Solar System orbit in the opposite direction. Now they saw that it has geysers that spew out material up to an altitude of several kilometers.

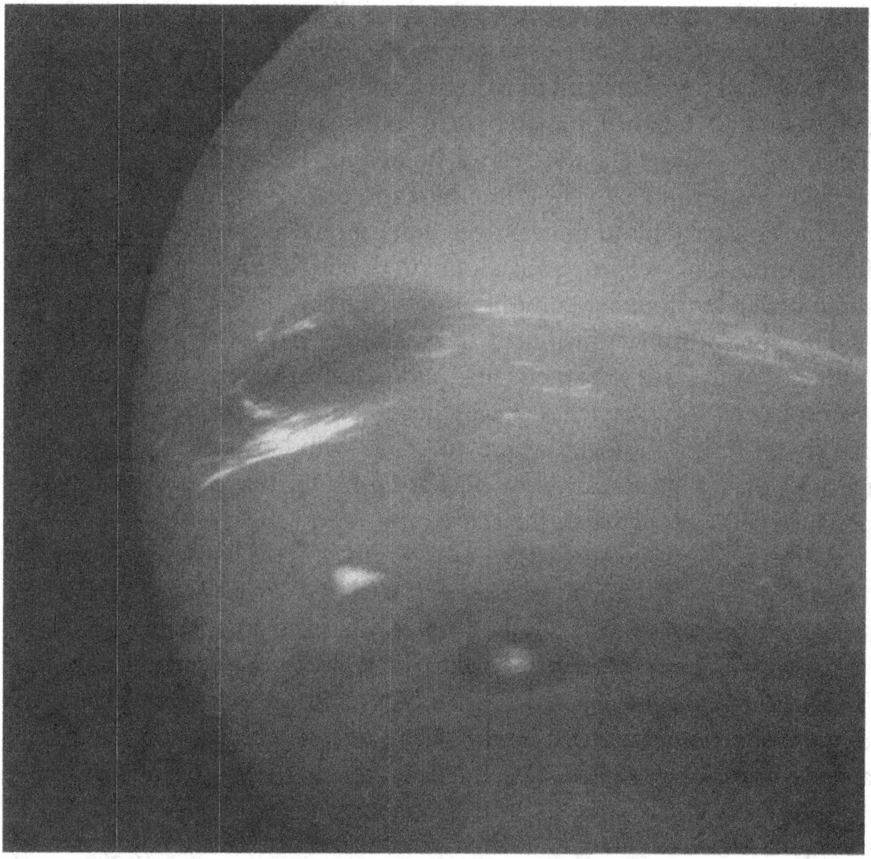

FIGURE 7.20 Neptune as seen by Voyager 2. [NASA]

Voyager 2 showed that Uranus and Neptune are very similar, and that both are quite different from Jupiter and Saturn. Although their atmospheres mostly consist of hydrogen and helium, they also contain relatively more methane, which causes the plain blue–green color of these planets. As we now think that their atmospheres hide icy or slushy mixes of water, ammonia and methane, Uranus and Neptune are therefore sometimes referred to as the "ice giants." Their thick atmospheres may even conceal huge water oceans.

In 1999 the University of California in Berkeley, USA, conducted laboratory experiments in which samples of methane were subjected to the conditions thought to be found deep inside Uranus and Neptune. The samples were pressurized to levels several hundred thousand times higher than on the Earth's surface, and heated to temperatures over 3,000 degrees Celsius (5,400 degrees Fahrenheit). The experiments produced diamond dust – both planets may have a kind of diamond rain, which slowly descends through the mix of water, methane and ammonia!

Voyagers 1 and 2 are still functioning and, respectively, are currently 14 billion and 11 billion kilometers (8.7 and 6.8 billion miles) from Earth. That makes Voyager 1 the record holder for the furthest space probe, signifying humanity's furthest leap into space.

The Voyagers' plutonium power units should provide enough energy for transmissions to Earth until about 2020, but NASA may decide to save some bucks by stopping communications earlier.

The last scientific mission of these famous probes is to measure the influence of the Sun's magnetic field and radiation beyond the realm of the main planets. Voyager 1 has already become the first probe to hit the so-called termination shock, where the solar wind of electrically charged particles that continuously blow outward from the Sun is slowed down by pressure from gas between the stars.

At this termination shock, the solar wind abruptly slows down from its average speed of 300 to 700 kilometers (190 to 440 miles) per second. In the same way that an initially steady flow of cars piles up at traffic lights, the collision with the interstellar gas makes the wind become denser and hotter. The higher density of charged particles also results in an increase in strength of the magnetic field carried by the solar particles. The decrease in velocity and the increase in magnetic field intensity were both detected by Voyager 1.

Like Pioneers 10 and 11, the Voyagers will continue to function as envoys of humanity long after their power runs out. Both Voyagers carry gold-plated copper disks with messages from Earth. If the probes ever get intercepted by our descendants or an extraterrestrial civilization who are able to decipher the instruction on how to play the disks, they will be able to hear the sounds of whales, the wind, the sea and rock 'n' roll music, and enjoy pictures of various locations on Earth, animals, people and everyday life on our world. Like a tourist brochure, the disks may invite aliens to contact or even visit us, if humanity still exists by then.

The Sun

By far most of the mass in the Solar System, 99.8 percent, is housed inside the Sun itself; the planets are mere aggregations of the dust and gas that were left over after the creation of this star. Fueled by nuclear fusion in its core, converting hydrogen into helium, the Sun provides heat and light to Earth. Without it, life as we know it would not be possible.

The Sun is very large and stands relatively close to Earth, so with specially constructed ground-based telescopes there is a lot of detail we can see: the swirling bubbles on the visible surface, the cooler and therefore darker sunspots, and the jets of extremely hot gas that are ejected into space.

Luckily for us, the Earth's magnetosphere diverts, and the atmosphere filters, a lot of the deadly radiation from the Sun. However, this also means that astronomers cannot detect everything that is happening on and inside our star from the ground. Another limitation is that from Earth we can only have a good look at the equatorial regions of the Sun; what is happening on its poles is hard to see.

In 1990, ESA and NASA therefore launched the Ulysses probe to climb above the plane in which all the planets orbit (ESA build the spacecraft, while NASA provided the launch with the Space Shuttle and the RTG to power Ulysses). Using a gravity swing from Jupiter, it went into a wide orbit over the Sun's poles.

Ulysses discovered that the solar wind, the stream of charged particles sent out by the Sun, actually has two components: a slow one emitted by the equatorial regions and a faster one blowing from the poles.

FIGURE 7.21 Artist impression of the launch of Ulysses from the Space Shuttle. [ESA]

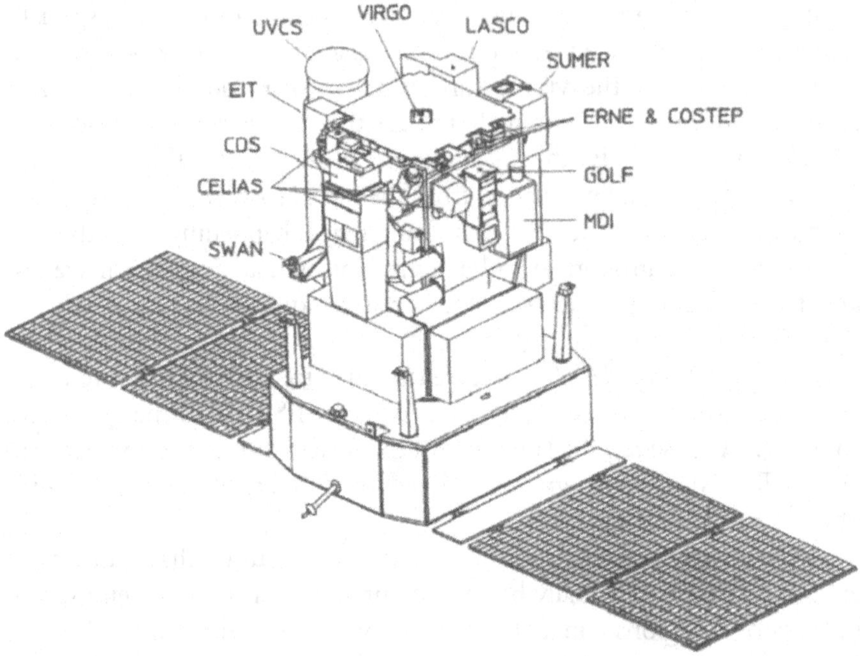

FIGURE 7.22 SOHO and the location of its various Sun-observing scientific instruments. [ESA]

In 1995 the Solar and Heliospheric Observatory mission, SOHO, was launched on board an Atlas IIAS rocket from Cape Canaveral Air Force Station. As a joint project of ESA and NASA, the Sun observation satellite was designed to continuously view the atmosphere, surface and interior of our local star. From its position far away from Earth, SOHO's view was never interrupted by our planet.

SOHO is orbiting the Sun in a very peculiar orbit. It is located at the so-called Earth–Sun L_1 Lagrangian point, where it can observe its target without its view ever being blocked by the Earth.

Lagrangian points are locations in space where the gravitational forces on, and the orbital motion of, a body balance each other. They were discovered by the Frenchman Louis Lagrange in 1772, when he described the mathematical "three body problem" of how a small object would orbit around two more massive bodies such as planets and stars. There are five Lagrangian points in the Sun–Earth system, L_1 to L_5, and L_1 is a point directly between the Earth and the Sun (such points also exist for the Earth–Moon system).

A spacecraft in this L_1 point always remains in the same position with respect to the Earth and the Sun, and therefore it provides SOHO with a continuous, uninterrupted view of the Sun and a direct line of communication with Earth. If SOHO was just orbiting the Earth, the Sun would often be eclipsed by the planet. If it was orbiting the Sun, it would be hard for ground stations on Earth to stay in constant contact with SOHO. The fixed L_1 position makes SOHO a stable observatory both with respect to the Sun and to Earth.

In April 1998 SOHO successfully completed its nominal mission, but it continued to work perfectly. On December 2, 2005, it celebrated its tenth anniversary in space, and at the time of writing it is still being used.

SOHO has allowed more than 3,200 researchers to make major advances in solar science. The mission revolutionized our ideas about the solar interior, the solar atmosphere and the dynamics of the solar wind. Major SOHO achievements include the detection of rivers of plasma beneath the surface, a complicated magnetic layer on the Sun's outside, and the first detection of flare-induced solar quakes ("earthquakes" on the Sun).

Moreover, it made many spectacular images and movies of eruptions called "solar flares." These are monumental explosions on the Sun that are

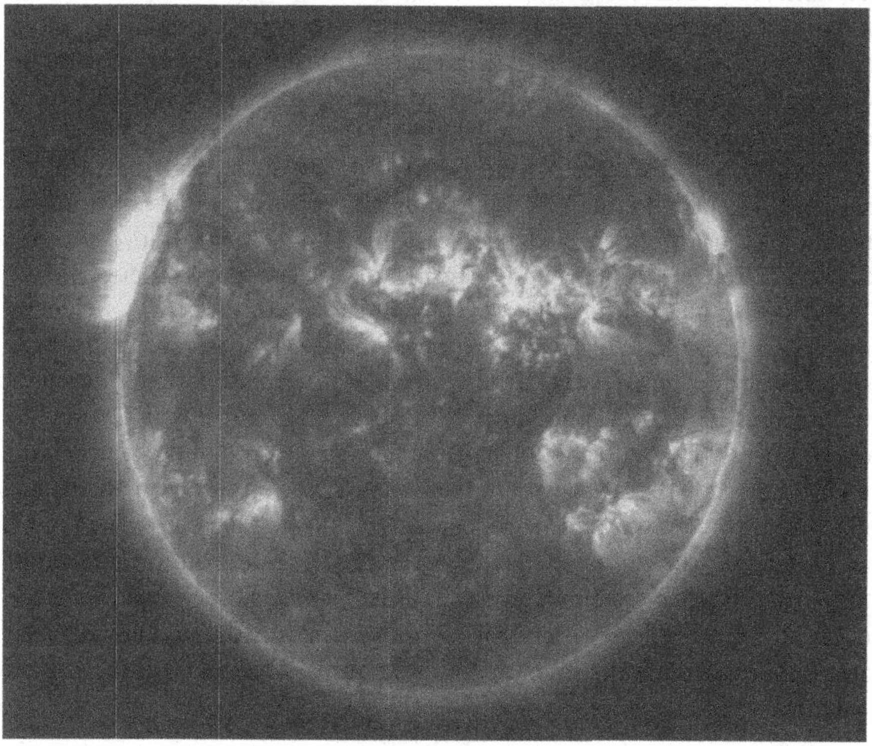

FIGURE 7.23 This is how ESA's SOHO spacecraft sees the Sun in ultraviolet, a wavelength invisible to the human eye. [ESA]

caused by sudden releases of magnetic energy, and emit as much energy as millions of atomic bombs. The flares accelerate the normal solar wind (the stream of energetic particles expelled by the Sun) to "storm force," and also trigger the expulsion of enormous amounts of gas from the Sun in what are called "coronal mass ejections."

SOHO can only see one half of the Sun at any time, but scientists used its data to develop methods to see what is happening on the other side. They found out that eruptions on the back of the Sun send ripples all the way around to the side visible to SOHO.

In this way we can get warnings about solar storms earlier than ever, even before they become visible to us. This is important, because severe geomagnetic storms (caused by the powerful burst of the solar wind hitting the Earth's magnetic field) can disrupt not only electronics on board satellites, but also radio communications and even the electrical power supply on Earth (and also cause beautiful aurora displays in the atmosphere).

SOHO also discovered more than 1,000 comets, which means that it has almost doubled the number of known comets. Comets are chunks of ice and dust that start to evaporate when they get close to the Sun, often developing beautiful long tails of expelled material. From its unique vantage point, SOHO is able to see comets grazing and sometimes even impacting the Sun.

Many comet discoveries have been made by amateurs using SOHO images on the Internet. SOHO and modern computer information technology has thus enabled people from all over the world, from the United States to Lithuania and Taiwan, to look for new comets.

Actual samples of the Sun were brought to Earth by NASA's Genesis mission. It captured solar wind particles in specially designed, ultra-pure silicon wafers. The Genesis capsule carrying the precious samples managed to return to Earth, but unfortunately crashed when its parachutes failed to open. Luckily NASA has still been able to salvage some of the wafers with their captured solar particles from the debris (more on this dramatic failure in Chapter 8, "Death of a spacecraft").

There are many other spacecraft observing the Sun from orbits around the Earth. Rather than exploring the Sun from a closer range, they are used as astronomical observatories above the distorting and filtering atmosphere to study the interaction of the Sun's radiation with the Earth's magnetic field.

Asteroids and Comets, the Building Blocks of the Solar System

Apart from the Sun, the nine planets and their many moons, the Solar System is full of small bodies called comets and asteroids. The first contain a lot of ice, the latter are rockier. On average, comets have lower bulk densities than asteroids.

However, due to fresh data from recent robotic explorers, the distinction between comets and asteroids is no longer very clear. Some asteroids are now suspected to be former comets on which all the ice has been evaporated by the Sun, leaving only the rocky core. Moreover, we have recently found out that rather than "dirty snowballs," as astronomers used to call them, comets are more asteroid-like "snowy dirtballs."

Fascinating as they are, asteroids and comets can be quite dangerous. When they hit planets they create havoc, as was impressively demonstrated when fragments of comet Shoemaker–Levy 9 crashed into Jupiter with the explosive force of several billions of atomic bombs. (Comets are named after the astronomers who discover them; this was the ninth comet found by Eugene and Carolyn Shoemaker and David Levy.) The geological history of the Earth is full of evidence of such collisions, like the famous Meteor Crater in Arizona.

We now think that an asteroid impact was at least partly responsible for the extinction of the dinosaurs. Some 65 million years ago, a large asteriod hit the Earth and filled the atmosphere with dust. The Sun's life-bringing light was blocked out for months. It became extremely cold and most of the plants died, breaking the food chain and causing the death of, first, the plant-eating dinosaurs and then the meat-eaters who preyed on them.

In this particular case the impact actually helped humanity, as the disappearance of the dinosaurs gave little furry mammals a chance to prosper, develop further and eventually evolve to become humans. However, the next time a large comet or asteroid comes crashing down we will surely not be so happy about it (just have a look at the movie *Deep Impact* or, much more funny, *Armageddon*, and you'll get the picture). The geological and fossil record shows that life has probably been partly wiped out of existence by asteroid and comet impacts many times in the Earth's history.

Attack from outer space

Between Mars and Jupiter there is a vast region full of small rocky bodies: the asteroid belt. The largest of these rocky worlds is Ceres, with a diameter of 950 kilometers (590 miles), but most of them are much smaller. Scientists estimate that there are several millions of them with diameters over a kilometer. The asteroid belt seems to be a failed planet, a junkyard full of scrap left over from the creation of the Solar System. The original dust particles sometimes merged into relatively big planetoids, but the resulting large chunks apparently never managed to combine to create a proper planet. This was mainly because they were continually being stirred up by the gravity of giant Jupiter.

Not all asteroids orbit in the belt. Some of them follow elliptical orbits that can bring them dangerously close to the Earth. As there are far more asteroids than comets coming close to the Earth, "planetary defense" should primarily focus on them. If we know well in advance that an asteroid is going to hit us, we may be able to land a probe with a large propulsion system on it and push it into a slightly different orbit. If applied on time, even a minute change in direction or speed can make it miss our planet.

However, before we attempt to do something like that, we need to know what the surface of such an asteroid looks like and what it is made of. The surface may, for instance, be too soft and unstable to withstand a lot of rocket thrust, or the uneven distribution of its mass may make it start to tumble if we push at the wrong place.

We should also be able to accurately predict the orbits of potentially dangerous asteroids. Former astronaut Russell Schweickart therefore advocates landing a radio beacon on 2004 MN4, an asteroid about 320 meters (1,050 feet) across that currently has a chance of 1 in 10,000 of hitting the Earth in 2036. With such a beacon attached, we could track this potential killer precisely and calculate its exact orbit to find out if we are on a collision course. If we are, we may be able to do something about it, but only if we find out in time. That is why Schweickart would like NASA to send a radio beacon probe as soon as possible.

However, scientists believe that ground-based observations of 2004 MN4 will be just as effective in fine-tuning the impact risk assessments. Until 2012 the asteroid is too close to the Sun for observation, but in 2013 tracking by telescopes on Earth should enable them to map its orbit more accurately. In 2029 it makes a close approach to Earth, and then we should be able to make the very refined trajectory determinations that will tell us whether we are in any real danger.

In the meantime, we have already learned a lot about asteroids. On its way to Jupiter, NASA's Galileo crossed the asteroid belt and took the opportunity to make the first close-up images of two large ones. Gaspra turned out to be a potato-shaped rock about 18 by 11 by 9 kilometers (11

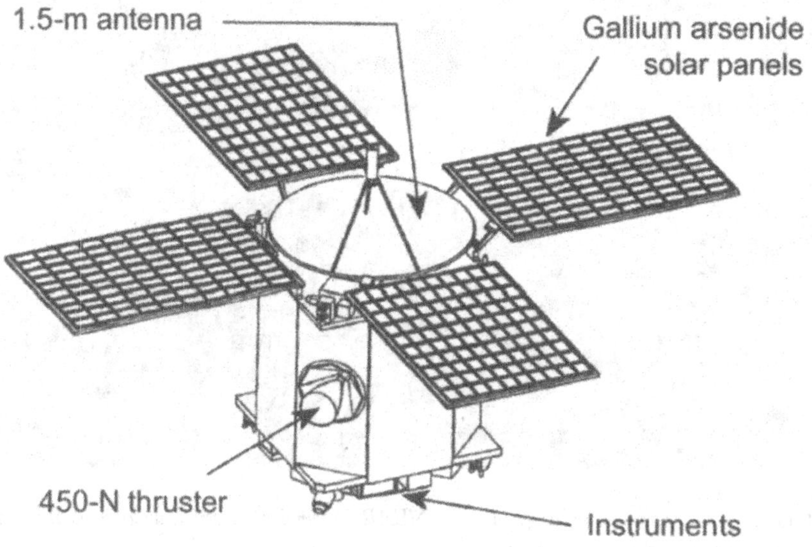

FIGURE 7.24 NASA's Near Earth Asteroid Rendezvous spacecraft. [NASA]

by 7 by 6 miles) in length. The second, Ida, was much larger. This asteroid – 55 by 24 by 20 kilometers (34 by 15 by 12 miles) in diameter – even turned out to have a little moon of its own, which was named Dactyl.

The first real asteroid mission was NASA's NEAR (Near Earth Asteroid Rendezvous). It became the first spacecraft ever to go into orbit around an asteroid. The prime target was Eros, an irregularly shaped asteroid about 13 by 13 by 33 kilometers (8 by 8 by 21 miles) in size.

The small, octagonal prism-shaped NEAR explorer was equipped with an X-ray/gamma-ray spectrometer and a near-infrared imaging spectrograph to find out what Eros is made of, a multispectral camera for detailed images, a laser altimeter to map elevations and depths on the surface, and a magnetometer to measure Eros's magnetic field. A radio science experiment using the NEAR radio-tracking system's Doppler shift would give some idea of the gravitational field of the elongated asteroid.

NEAR, later renamed NEAR Shoemaker in honor of the famous deceased asteroid and meteorite scientist, Eugene Shoemaker, was launched in February 1996, flew within 1,200 kilometers (750 miles) of another asteroid called Mathilde in June 1997, swept past Earth in January the following year, and then finally got into orbit around Eros on February 14, 2000, the aptly named St Valentine's Day.

Even though Eros is a quite big asteroid, it is still quite a small object and therefore has a very weak gravitational field. As a result, NEAR's orbit around it was very slow and very low; it circled the asteroid at only a

FIGURE 7.25 Eros was investigated by the NEAR Shoemaker spacecraft, which made this picture of the potato-shaped asteroid. [NASA]

couple of kilometers per hour and got as close as 24 kilometers (15 miles) to the surface. This naturally enabled it to do some great science at close distance, and even enabled it to softly touch down on the surface of Eros at the end of its life (more on this in "NEAR Eros").

The next asteroid mission, Deep Space 1, was launched in October 1998. Its primary task was to test new technology for future interplanetary spacecraft, in particular ion propulsion. But just like the later SMART 1 lunar satellite of ESA, NASA's Deep Space 1 also had a scientific goal.

It flew past the 3-kilometer (2-mile) diameter asteroid 9969 Braille on July 28, 1999, at a distance of less than 10 kilometers (6 miles). The spacecraft then moved on and passed the nucleus of comet 19P/Borrelly at a distance of 2,200 kilometers (1,400 miles) in September 2001.

Japan is also joining the asteroid hunt. The ion-propelled MUSES-C probe was launched in 2003, and was renamed Hayabusa once it was put in space. After an Earth swingby on May 19, 2004, the spacecraft made a rendezvous with near-Earth asteroid 25143 Itokawa in September 2005. Asteroids are named by their discoverers, and appropriately this one was named after Hideo Itokawa, the father of rocket science in Japan.

In November Hayabusa released a 600-gram (1.3-pound) lander named Minerva. The little probe was to have photographed the asteroid's surface and recorded temperatures there, but unfortunately contact was lost soon after the lander detached from its mothership.

Nevertheless, the orbiter itself continued to operate reasonably well. For months it made detailed observations while circling the asteroid, then moved in to land on it for a brief period – a bit like a bird of prey swooping down on its victim, but in slow-motion. (In fact, the spacecraft's Japanese name means "peregrine falcon.") Hayabusa was planned to obtain a small sample by shooting a kind of bullet into the surface, then collect some of the expelled debris.

In November 2005 it made two attempts to do this. The first time it touched down, bounced up once, spent 39 minutes resting on the surface and then launched back up in orbit again. Unfortunately it did not shoot its "gun," and thus did not manage to gather any surface material. A week later the second attempt may have been more successful. JAXA, the Japanese Space Agency, said that Hayabusa then probably touched down for a few seconds on the asteroid, and this time the pyrotechnical sample system may have worked successfully.

At the time of writing, Hayabusa is on its way back to Earth. We will not know whether it indeed managed to obtain some powder from the asteroid's surface until its re-entry capsule lands in 2010, hopefully with a few grams of pristine asteroid material. If so, it will be the first sample of an

FIGURE 7.26 Beyond Pluto there are other large planet-like objects that have not been visited by any spacecraft, such as the recently discovered Quaoar and Sedna.

asteroid we can investigate in a laboratory, other than the meteorites found on Earth which we know originate from asteroids. (However, at the time of writing the spacecraft is experiencing major problems with its attitude control, which has already moved the return date from 2007 to 2010.)

NASA's New Horizons Pluto Kuiper Belt Flyby mission was launched in January 2006. Rather than seeking out the asteroids in the inner Solar System, this mission is going much further out. It will first pass Jupiter for a velocity-boosting gravity assist maneuver early in 2007, then reach Pluto and its largest moon Charon in 2015. It will make detailed images and measurements as it passes by at high velocity. After that, the spacecraft is planned to fly on to investigate some of the icy planetoids (large asteroids) that orbit the Sun beyond Neptune, and include Pluto, in a region called the Kuiper Belt.

Earth-based telescopes have already found some planet-like objects beyond Pluto, like Quaoar, a planetoid one-third the diameter of the Moon. The largest one confirmed is Sedna, which has an estimated diameter of about 75 percent of that of Pluto.

However, astronomers have perhaps found a planetoid that is even larger than Pluto. It appears to have a moon, and observations with NASA's Spitzer space telescope have shown that there is methane-ice on its surface. Until it is confirmed, the object is named "2003 UB_{313}."

As usual, for probes flying this deep into space, the New Horizons mission will incorporate an RTG with radioactive material for electrical power supply. However, there is currently a shortage of the needed plutonium-238, due to a security-related shutdown of the US Department of Energy lab that processes this radioactive material. This means that the probe did not get all the plutonium originally planned.

Whether New Horizons will have sufficient amounts of energy to continue its mission after the Pluto flyby is therefore uncertain. Postponing the launch until all the required plutonium was available would have added three to five years to the probe's transit time and millions of dollars to the mission's cost.

Later in 2006, NASA plans to launch a deep space probe named Dawn. Using ion propulsion, as pioneered by Deep Space 1, Dawn will travel for nine years to reach the two most massive asteroids known, Vesta and Ceres. It will first go into orbit around Vesta and stay there for about nine months. Then, using its solar-electric engine, it will depart and journey further out to reach an orbit around Ceres. Its investigations there are also planned to take about nine months.

Both of the small planets to be visited are located in the main asteroid belt between Mars and Jupiter, but they are very different from each other.

Observations from Earth have revealed that the surface of Ceres is probably rather "primitive," meaning that it has not been altered much by geological processes. It seems to be covered by a layer of dry clay that contains water-bearing minerals, and it possibly has a very weak atmosphere and frost.

Vesta was more active during its early life; it has been resurfaced by basaltic lava flows and may have been (partly) covered by a magma ocean. Wrapped in a layer of basaltic dust, it is also much drier than Ceres.

Vesta has been pounded by smaller space rocks many times during its life, the debris of which have sometimes reached Earth. We know that some of the meteorites found on our planet came from Vesta because they have the exact same composition as the asteroid's surface.

We do not know whether we already have pieces of Ceres in our meteorite collections, because we have not yet been able to determine the composition of Ceres below its clay cover layer. The only way to find out what Ceres is made of is to pay the asteroid a visit, as Dawn will do.

The Dawn mission will investigate the main attributes of the two asteroids, such as their shape, size, mass, composition, density, their magnetic field and the numbers and sizes of craters on their surface. The most important issue that the mission will address is the role of size and water in the evolution of planets.

Thawing ice balls

The Solar System is surrounded by an immense spherical cloud of comets called the Oort cloud (Oort was a Dutch astronomer who first hypothesized its existence), enveloping the Solar System far beyond the orbit of Neptune. The cloud extends to perhaps around two light-years, or 20 million times a million kilometers (12 million times a million miles) – half the distance to the next nearest star.

The comets are very loosely bound by the weak solar gravity at such distances, and can thus easily be disturbed by the passage of other stars. Their originally more or less circular orbits far from the Sun can then be readily changed, hurtling them into the heart of the Solar System. Following highly elliptical orbits these deep frozen, icy objects come closer to the Sun and partly evaporate, leaving beautiful tails of gas and dust behind.

Comets are thought to be fossil left-overs from the formation of the Solar System, 4.6 billion years ago. They should contain the original material from which the planets, including the Earth, were made. Some scientists also theorize that a proportion of the water on our planet was delivered by comet impacts, and that maybe even the basic ingredients of life were deposited by them. It is no wonder that comets have become the target of many recent space probes.

The first ever mission to a comet only took place in 1985, when NASA's International Cometary Explorer, ICE, passed through the tail of comet Giacobini–Zinner at a distance of 26,550 kilometers (16,500 miles).

ICE had originally started its life as the International Sun–Earth Explorer 3 satellite, and had been studying the interactions between the Sun and the Earth's magnetic field. With the help of some gravity assist lunar flybys, the spacecraft was sent on its way to the comet and renamed. In its new mission as ICE it studied the effect of the Sun's radiation on the gas in the tail of the comet.

The first mission specifically designed to study a comet from closeby, Giotto, was also ESA's first ever interplanetary spacecraft. The probe was designed to encounter the famous comet Halley in 1986. Hurtling through the gas and dust surrounding the comet, it got as close as 596 kilometers (370 miles).

The most difficult problem was how to ensure that Giotto survived long enough while flying through comet debris at a relative speed of 245,000 kilometers (152,000 miles) per hour. At this velocity, you could cross the Atlantic Ocean in only 11 minutes! To stop even a small, 0.1-gram dust particle from the comet at this speed would require a solid aluminum shield at least 8 centimeters (3 inches) thick. For Giotto, this would mean that the shield would weigh over 600 kilograms (1,300 pounds) – much too high for the small spacecraft.

Instead the designers chose to install two protective layers. The front, 1-millimeter (0.04-inch) thick sheet of aluminum would vaporize all but the largest of the incoming dust particles. A 12-millimeter (0.47-inch) thick sheet of Kevlar placed 23 centimeters (9 inches) behind the first would then absorb any debris that pierced the front barrier.

Together the sheets could withstand impacts from particles up to 1 gram traveling 50 times faster than a bullet, and this is achieved by a combined total mass that is far less than the mass of a single shield, which offers less protection. The combined sheets are called a Whipple shield, named after the American astronomer Fred Whipple who thought of it as early as 1947.

The Whipple shield worked very well. Fourteen seconds before its closest approach, Giotto was hit by a relatively large particle that slightly changed its attitude, but the sturdy little probe continued its journey intact.

The onboard experiments returned a wealth of scientific data, the most important being the images of the comet's nucleus. Giotto discovered that the heart of Halley's comet is a 16-kilometer-long (10-mile-long) peanut-shaped, coal-black body made up of dust and ice. Heated up by the Sun, fountains of gas (from the ice) and dust particles could be seen spewing out of the comet, creating a foggy halo around it and a long tail behind it.

Giotto survived the hazardous encounter rather well; most experiments suffered only minor damage and remained operational. However, the mass spectrometers, one sensor on the dust detector, a sensor on a plasma analyzer and also the multicolor camera were left permanently inoperable.

The spacecraft was therefore sent on an extended mission to another comet. It flew by comet P/Grigg–Skjellerup in July 1992, at a distance of only about 200 kilometers (124 miles). Sadly, no images could be made because of the broken camera.

Giotto was not the only probe to take advantage of Halley's visit. Japan sent two probes, Sakigake an Suisei, and the Russians sent their two Vega probes that first flew to Venus then visited the comet. However, none of them came as close as Giotto.

Encouraged by the success of Giotto, ESA embarked on an even more ambitious mission when it launched the Rosetta comet explorer in March 2004. It will take this large spacecraft more than 10 years to reach Comet 67P/Churyumov–Gerasimenko. If everything goes right, it will go in orbit around the comet in 2014 and start the first long-term exploration of a comet at close quarters.

Just as the famous Rosetta Stone enabled us to decipher Egyptian hieroglyphs, so we hope that the Rosetta spacecraft will make us understand what comets are made of and thereby what the early Solar System looked like.

Rosetta will even detach a 100-kilogram (220-pound) lander called "Philae" – about the size of a large TV. (Philae is the Nile island on which an obelisk was found that helped French historian Jean-François Champollion to decipher the hieroglyphs of the Rosetta Stone, and thereby unlock the secrets of the civilization of ancient Egypt.)

To prevent Philae from bouncing off due to the extremely low gravity of the small and low-mass comet, the lander will secure itself by firing a harpoon into it. The design of this harpoon has been something of a gamble, as no one really knows the properties of the comet material it will hit. Maybe it's so loose that the harpoon will not get a foothold. Or it might be rock-hard ice and the harpoon will simply bounce off. Scientists and engineers made their best guess and hope to find a reasonably soft but not too loose surface.

Another comet mission recently launched was NASA's Contour (Comet Nucleus Tour). It was supposed to fly past two or maybe three comets. Sadly, after ignition of the solid propellant rocket stage that was to push the probe out of Earth orbit, contact was lost. Through telescopes, observers saw three separate objects where Contour should have been; apparently the spacecraft and upper stage had fallen apart.

An investigation board later concluded that probably the exhaust from the rocket booster heated up the spacecraft much more than expected; the structure could not handle the high temperatures and broke up. This failure once again showed that there is still nothing routine about space exploration.

A more successful mission was NASA's Stardust. It flew past comet 81P/Wild-2 in January 2004, catching material released by the active comet using a special substance called "aerogel," which slows down and captures the particles without damaging them. A return capsule with this precious cargo landed on a lakebed in Utah on January 15, 2006, giving scientists direct access to pristine comet material for the first time.

A very short but truly spectacular interplanetary mission, Deep Impact, was launched in January 2005. The probe reached its target, comet

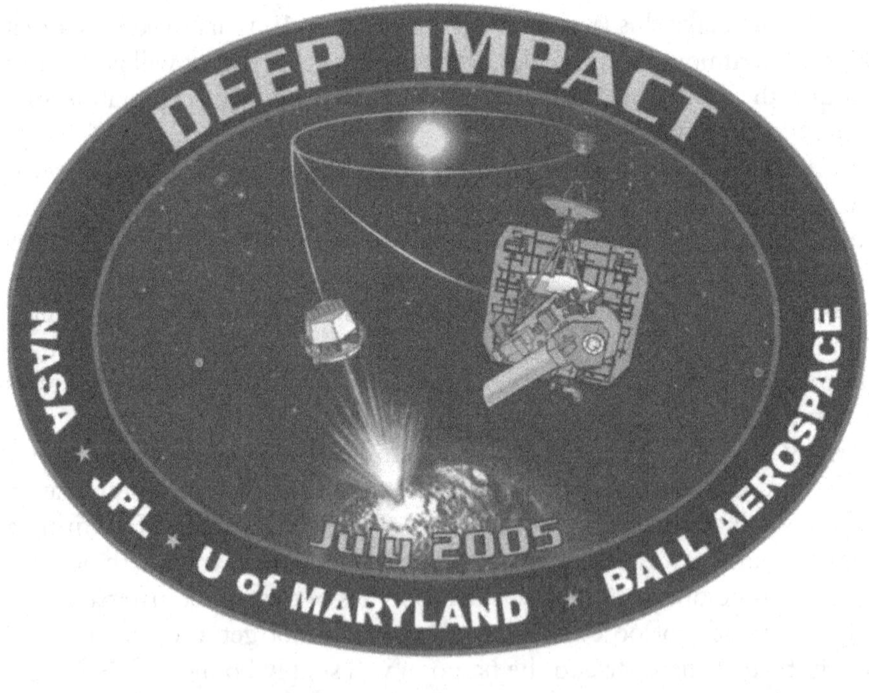

FIGURE 7.27 This mission logo shows the Deep Impact mothership Flyby and the smaller Impactor that separated to hit comet Temple 1. [NASA]

Temple 1, on July 4 of the same year. There the spacecraft split in two, as the "Flyby" mothership released a smaller, 370-kilogram (820-pound) "Impactor" probe in front of the onrushing comet. Several maneuvers were needed to ensure that the probe would hit at the right spot.

Over the Internet, Impactor's camera showed viewers live pictures of the comet rapidly approaching. The images showed impressive craters and ridges on Temple 1's surface, getting larger and larger. The final picture was made about 30 kilometers (19 miles) from the surface, showing details as small as 4 meters (13 feet) across.

Then the stream of pictures suddenly ended, as Impactor smashed violently into the comet's icy body at a relative speed of 37,000 kilometers (23,000 miles) per hour. The impact of the washing machine sized probe released as much energy as the explosion of 4,500 kilograms (10,000 pounds) of TNT!

Subsequently, the Flyby mothership zoomed by the interplanetary crash site at a safe distance of 500 kilometers (300 miles) and monitored the impact results. Its first images showed a huge flash of light created by the demise of the Impactor probe. The illumination lasted less than a second, but like a giant photoflash it provided an excellent light source for the two cameras on Flyby.

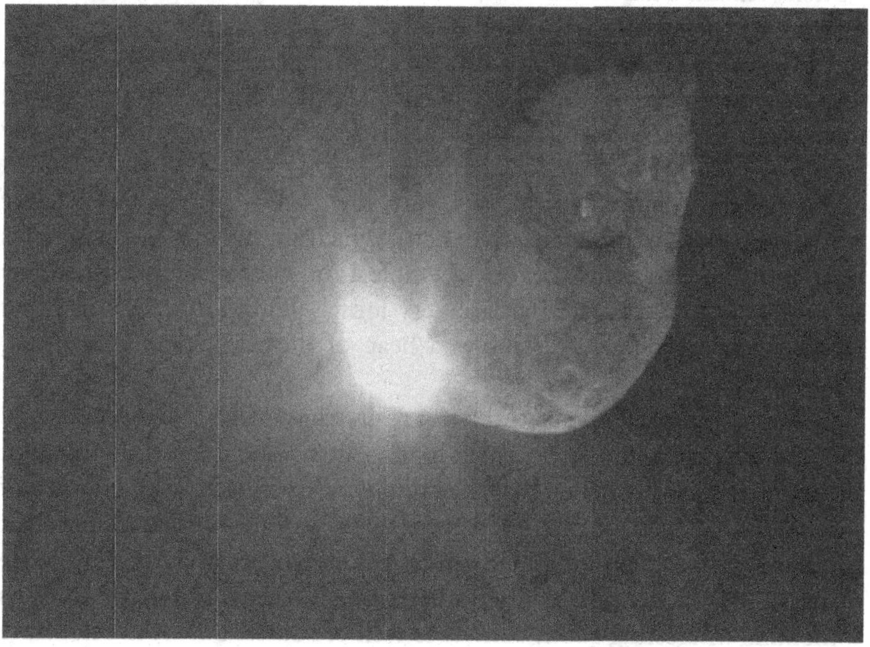

FIGURE 7.28 Huge amounts of gas and dust are spewing out of Temple1 where Impactor has hit. This picture was made by the Flyby spacecraft shortly after the impact. [NASA]

Their images showed massive amounts of dust and gas spewing out of the impact location, at an incredible 5 kilometers (3 miles) per second. Later analyses indicated that the collision left a crater close to 250 meters (820 feet) in diameter.

The observation of the impact was a nice example of international cooperation in space exploration. Apart from Flyby, many observatories on Earth, space telescopes such as Hubble and XMM Newton, and even the Rosetta spacecraft were used to watch the spectacle with as many types of instruments as possible.

The Hubble Space Telescope monitored the expansion of the plume of dust jetting out of the comet, which is about half the size of the island of Manhattan in New York. One hour after impact it showed that the debris had already extended 720 kilometers (450 miles) from the comet nucleus. Some 18 hours later it was looking at a cloud that had grown to just less than 3,200 kilometers (2,000 miles) across.

By analyzing the spectrum of the sunlight that was reflected by the cometary materials, scientists were able to determine its composition. The first analyses showed that a lot of hot water vapor and hot carbon dioxide vapor was coming out – the initial heated material from the impact. Almost all the material ejected from the crater was in the form of tiny dust particles smaller than 10 microns, which is really tiny. The particles were a mixture of rocky dust and volatile solids.

Comparison of the pre-impact to the post-impact spectra showed a 10-fold increase in the amount of water and carbon dioxide, but a 20-fold increase in the amount of organic material, possibly excavated from the comet's interior.

The overall density found was very low, only 0.6 gram per cubic centimeter (0.02 pound per cubic inch), which means that the material is light, fluffy and extremely porous. Temple 1 must therefore consist largely of extremely fine particles that are very loosely bound together. In other words, the comet more resembles a heap of snowdrift than a compact, hard snowball.

Before the impact, scientists did not really know what to expect. Would the comet's consistency be rock hard, soft as snow or something in between? It was anticipated that the crash would result in large amounts of material being thrown out. However, there was also a slim chance that Impactor would quietly disappear into a fluffy comet body of very loose material, or shoot out at the other end. That would have told us something interesting about comets as well of course, but would not have allowed the analysis of the comet's inner composition.

Despite its complexity, nearly everything on the mission went better than expected. A Russian astrologer nevertheless complained that the impact had upset her horoscopes and tried to sue NASA for 300 million dollars!

After analysis of the data, Deep Impact will have taught us much about what comets consist of, how fragile they are and how large craters are formed on other planets and moons. Maybe such data will also help us to protect ourselves against possible comets heading for Earth.

Never Finished

Now, in the first decade of the twenty-first century, we have visited all of the large planets of the Solar System. Only the last of the original nine has still been left alone, but as scientists are regarding Pluto more and more as a rather large, icy planetoid, we may be excused for this. Eight out of nine isn't bad. Moreover, NASA's New Horizons spacecraft is on its way for a fast flyby of Pluto in mid-2015.

We have also visited and even landed on some of the most interesting moons, asteroids and comets, and have sent out probes to study the Sun itself. However, to conclude from this that our exploration work is nearly done would be a mistake.

First, of all the hundreds of moons and hundreds of thousands of large asteroids and comets, we have visited only a very, very small fraction. Most of them we know very little about, and many have not even been discovered yet. You only have to take a look at the diversity of the moons of a single planet such as Jupiter or Saturn to get an idea of the number of objects waiting to be discovered.

Second, the fact that we have made a couple of visits to a planet or moon does not mean we know everything there is to know about it, or even enough to be reasonably content. If an alien probe were to land in the Sahara, it might signal to its masters that the surface of the Earth is hot and barren, and that liquid water and life are sparse. It would take a lot of landers and satellites for alien explorers to fully comprehend the diversity of our planet.

Mercury, Uranus and Neptune have so far only been visited very briefly by flyby spacecraft. Moreover, we have never put anything in orbit around a moon of any other planet than our own.

We do not yet even know enough about the Earth to truly understand its past or to predict the future of its climate, even though we live on it and have been exploring it by foot, with horses, boats, submarines, balloons, airplanes, cars and satellites for thousands of years. We have employed unimaginable numbers and diversities of scientific instruments to answer the questions about our home planet, only to find more questions that have even resulted in entirely new fields of science.

Most of the Earth's oceans and seas, which comprise the greater part of its surface area, have yet to be explored. How can we then ever assume to find out all there is to know about the rest of the Solar System by launching some tens, hundreds or even thousands of interplanetary missions?

Most aspects of the worlds we have explored have not yet been investigated. New scientific instruments with new capabilities are needed to answer remaining and new questions, and novel missions are required to take those instruments to their destinations.

A small sample of the big questions about the Solar System that remain indicates how much there is still to do:

1. Did life ever exist on Mars or any other planet or moon?
2. What are the characteristics required of a planet or moon to have life developing on its surface?
3. Does extraterrestrial life exist anywhere in the Solar System today?
4. How did the Sun and planets form, and what does that mean for the probability of Earth-like planets around other stars?
5. What has caused the atmospheres of Venus and Mars to be so vastly different from that of Earth?
6. Are there really large rings of comets and icy planetoids orbiting the Sun far beyond Neptune?
7. Do other planets and moons potentially have sufficient appropriate resources for people to live there in the future?

New space missions will search for life on Mars, past and present, and will bring samples of the red planet's surface back to Earth for detailed study. Landers will be sent to drill or melt through the thick layers of ice that are hiding possible oceans of water on Europa and other moons. Special probes will further explore Titan and find out how it got its strange atmosphere and surface. Long-distance spacecraft will explore Uranus, Neptune and their moons, and visit Pluto and other smaller bodies at the edge of the Solar System. They will also continue further out to ascertain where the Sun's influence and family of planetary bodies really end.

One day we may even be able send probes out to other stars. Once we can do that, billions of other Solar Systems await our exploration. Just think about the amount of investigation still left to do! Part of the beauty of space exploration is that it never finishes.

8

DEATH OF A SPACECRAFT

SPACE museums are a bit like natural history museums, full of skeletons and stuffed animals: nothing is alive, and if something moves it is usually merely an animated model of the real thing. The exhibition subjects are sometimes shown in simulations of their natural environments, but as hard as it is to capture the wildness of a jungle inside a museum building, it is even harder to simulate an alien world using wood and plaster.

Moreover, space museums can usually only show scale models or test hardware built during the development phases; the real stuff is out there, in space. As real heroes should, most robotic space explorers do not end up in an exposition but finish their operational life on duty.

Those that do not perish during launch failures or because of design errors mostly find their end due to empty propellant tanks or failing batteries. Some spacecraft are deactivated on purpose because they have outlived their usefulness; an explorer that ends up in empty space after a successful flyby of a comet or planet is often of no further value. Sometimes politicians pull the plug because the possible extra science resulting from an extended mission is not deemed worth the additional money required.

Let's have a look at some examples of famous endings and remarkable recoveries out there in deep space.

Down to Earth

Earth-orbiting spacecraft end their lives in most inglorious ways. Crippled satellites in low Earth orbits, below an altitude of 1,000 kilometers (622 miles), are slowly de-orbited by the small but continuous braking effect of the thin atmosphere that is still present up there. Without propellant for their rocket engines or electrical power to activate them, their orbital velocity and altitude cannot be maintained.

Spiraling down, the atmosphere gets ever denser and the drag increases, until the Earth's airy grip gets a firm hold. Then the poor satellites have a very rapid descent. Colliding with thicker air at 25 times the speed of sound, they burn up and often evaporate completely. If anything is left, it usually ends up at the bottom of the ocean.

Sometimes old satellites are even actively de-orbited on purpose, to remove them to ensure that they do not end up being space junk that endangers other spacecraft (imagine a satellite with a velocity of nearly 8 kilometers (5 miles) per second hitting another spacecraft coming from the opposite direction at the same speed . . .). They can be sent to their fiery end by using their last bit of propellant to slow them down and bring them back to Earth.

However, satellites in higher orbits usually remain there virtually forever. There is no atmosphere to brake their speed, and they have insufficient propellant to push them back down. Sometimes they are expelled into graveyard orbits full of retired satellites, to clear their slot in useful orbits for new spacecraft. The Earth is surrounded by thousands of such dead spacecraft, which often fall apart after a while and become even more space junk.

For large satellites in low Earth orbits such as space stations, active and controlled de-orbiting is vital; they are so big that large pieces do not burn up completely and thus reach the ground. Such spacecraft must be pushed back into the atmosphere at a precisely timed moment, or pieces could fall down in inhabited areas.

The de-orbiting of the giant Mir space station was truly spectacular, with the huge assembly of coupled modules braking down into individual elements that each tracked long arcs of fire across the sky. Big pieces reached the Earth's surface, but fell down harmlessly into the ocean thanks to the excellent planning of the Russian ground control teams.

However, not all spacecraft falling back to Earth burn away or finish up at the bottom of the sea. Some interplanetary robotic explorers do return to Earth more or less intact. As described before, in the 1970s the Soviet Union launched a series of soft-landing Luna Moon probes that drilled up some lunar soil, then launched their cargo back to Earth. Small re-entry capsules with parachutes then landed the precious material in Soviet territory.

More recently, NASA's Genesis mission returned a capsule with samples of the Sun. Genesis was launched in August 2001 to capture solar wind particles expelled by our star with the aid of ultra-pure silicon wafers. These wafers were designed to catch, slow down and contain the solar particles without damaging them. Scientists are very interested in these particles, because they can give vital information on the composition of the Sun, and thus on the origins of the Solar System.

The wafers were very fragile, and the mission's engineers feared that even a relatively soft landing under parachutes would damage them. They came up with a plan to hire stunt pilots to snatch the capsule out of the air with helicopters, each equipped with a special hook. The hook would catch the parachute lines, after which the helicopter could gently put the Genesis capsule on the ground. More or less the same principle had been used in the 1960s for Corona spy satellite capsules secretly returning with film full of pictures of Soviet territory.

The Genesis capsule returned to Earth at exactly the planned time and location, but unfortunately its parachutes failed to open. Horrified scientists and engineers watched the sample return capsule tumble down and smack into Utah soil at 320 kilometers (200 miles) per hour.

Lying on the ground, cracked open, the spacecraft somewhat resembled a crashed flying saucer. Most of the wafers were all smashed up and contaminated by pieces of spacecraft and Utah dust. Nevertheless, NASA has been able to salvage some of the captured solar particles, so the mission is not a complete failure.

Investigating the accident, NASA and Genesis prime contractor Lockheed Martin discovered that a simple blunder had prevented the parachutes from opening. A gravity switch, meant to trigger the deployment of a drogue chute to stabilize and decelerate the capsule, had been placed upside down. As the switch never activated the drogue, the main parachutes could not be deployed.

NASA had another spacecraft built by Lockheed Martin heading home:

the Stardust spacecraft with its return capsule full of comet dust particles emitted by other stars. Stardust used the same type of capsule and parachute subsystem as Genesis. However, this time the essential gravity switch had been properly installed and the precious cargo landed softly and safely in January 2006.

In a Blaze of Glory

Interplanetary spacecraft often end their useful lives substantially more honorably than Earth-orbiting satellites. NASA's Pathfinder Marslander, for instance, was renamed Sagan Memorial Station, after the deceased scientist and space promoter Carl Sagan. It thus lives on as a fitting monument on another planet.

The camera of the automatic lunar lander Surveyor 3 was returned to Earth by the Apollo 12 crew that landed close to it. Amazingly, engineers investigating it found it contained a streptococcus bacterium that had survived over 2.5 years on the Moon! It had probably been sneezed into the camera by a technician with a cold. The famous camera can now be seen in the Smithsonian Air and Space Museum in Washington, DC.

For some spacecraft, the arrival at their destination means that their (planned) end follows soon. Such was the fate of the Huygens probe that parachuted down into Titan's atmosphere, or the Ranger probes that were sent to crash on the Moon in order to make low-altitude photographs.

However, others are sent on a spectacular finale even though their main mission did not require it. The first deliberately crashed operating spacecraft was NASA's Magellan Venus radar mapper. As described before, in 1994 sustained aerobraking caused its mission to end with a dramatic plunge to the planet's surface. The purpose was to gather data on Venus's atmosphere.

Described in the previous chapter, NASA's Lunar Prospector was a small moon orbiter launched in 1998. It had a very successful mission of a year, during which it mapped the entire Moon and made detailed investigations of its surface composition. The probe then got a life extension of half a year for more detailed examinations from a lower orbit, circling at an altitude of only 30 kilometers (19 miles) above the lunar surface.

Nevertheless, in July 1999 all its work was done. Instead of being abandoned in lunar orbit, the little probe was given one more task that would abruptly end its mission: it was to crash on the south pole of the Moon.

The impact was hoped to eject water-ice thought to be hidden in the deep south pole craters, so that it could be detected by telescopes on Earth. Lunar Prospector hit right on the mark, but unfortunately no traces of water were seen by Earth-based observers. Whether the spacecraft went down into the wrong crater or whether there is just no water-ice on the Moon is unknown. Future lunar explorers will have to deal with this question.

Another successful spacecraft that ended its mission in a sudden but useful way was Galileo. Deployed from a Space Shuttle in 1989, it reached Jupiter in December 1995 and conducted detailed investigations of the largest planet in the Solar System. It observed its atmosphere, magnetosphere and its moons and sent us beautiful pictures of this far away world.

FIGURE 8.1 The Galileo spacecraft on the ground. Even though the umbrella antenna on top (seen folded here) did not deploy properly, Galileo's mission was a huge success. [NASA]

Although the parabolic umbrella antenna failed to open after launch, workarounds were developed to have Galileo send its data via a less powerful antenna with a lower gain. New data compression techniques made it possible for the spacecraft to send its pictures in a smarter way than originally foreseen, such as leaving out all the black space areas in the images. Even without the use of its main antenna, all the mission's science objectives were eventually met.

After eight years of cruising the Jovian system, Galileo's propellant was finally depleted. The spacecraft had to be destroyed, because of the possibility that Jupiter's uneven gravitational field would gradually change its orbit in such a way that it would crash on the moon Europa. If that happened, it might biologically contaminate it, and disturb future missions searching for indigenous life. Europa is considered to be one of the places in our Solar System where life may possibly have evolved, as its icy surface may be hiding a global ocean of liquid water, possibly containing organic material.

Galileo was therefore maneuvered to burn up in Jupiter's atmosphere on September 21, 2003. At 48 kilometers (30 miles) per second it dived down into the swirling clouds. The spacecraft continued transmitting data at least until it passed behind the limb of Jupiter, at which point it was 9,283 kilometers (5,768 miles) above the 1-bar atmospheric layer (because it is a gas planet that, as far as we know, has no solid surface, you cannot measure altitude above ground at Jupiter).

Engineers had expected Galileo's electronics to be disrupted as a result of the high radiation environment so close to Jupiter, but the venerable spacecraft worked until the end. The result was some data on pressures and densities in Jupiter's atmosphere.

The Cassini spacecraft that is currently orbiting Saturn may await a similar fate. Once the real mission is over, scientist can take a lot more risk with their expensive probe, and even send it on kamikaze missions.

Other than diving into Saturn, it may however also be steered into orbit around Titan. Moving it in closer and closer, it would give scientists another opportunity to get a good look at the moon's atmosphere. Cassini's mission would then end with a plunge into the murky Titan atmosphere and a fiery finale by burning up.

Another possibility is a flight across the rings of Saturn, probably through the open area called the Cassini Division. The rings are not very thick, but the chance that the spacecraft would hit a piece of rock or ice is relatively high, even in the nearly empty division. If Cassini survives, it may then still be sent into the atmosphere of Saturn or Titan.

Alternatively, Cassini may await a completely new mission: if sufficient amounts of propellant have remained, it could be boosted away from the Saturn system to meet a comet or asteroid.

SAVING SOHO

In June 1998 contact with the Solar and Heliospheric Observatory (SOHO) – the joint project of ESA and NASA for studying the Sun – was suddenly lost. In its stable Lagrangian point, the spacecraft had apparently somehow gotten into a position that prevented its antennas being pointed to Earth. Later investigations showed what had happened.

SOHO's attitude control uses three gyroscope units to tell its attitude with respect to the Sun and Earth: Gyros A, B and C. Gyro A is meant to be used in "safe mode," the mode of operations to which the spacecraft automatically turns when something is wrong. Gyro B is used to detect faults in the attitude control, and Gyro C is for routine attitude control.

The problems started with an erroneous software patch that was sent to SOHO and installed on its onboard computer. The new computer program accidentally allowed the critically important Gyro A to be deactivated, minimizing "wear and tear" in order to conserve its operational life.

After a routine maneuver using all gyros, SOHO's computer detected that Gyro B was giving false readings. The computer thus automatically put SOHO in safe mode and switched attitude control from the normally used Gyro C to the "trouble shooting" Gyro A. The error in Gyro B was then corrected and SOHO tried to re-align it with the readings from Gyro A.

However, Gyro A had just been deactivated by a command from the erroneous software. Now Gyro B implied that the spacecraft was rotating, while Gyro A implied that it was standing still. By firing its thrusters to change the roll velocity, SOHO desperately tried to move itself so that both gyros would indicate the same speed and thus be re-aligned.

Unfortunately this could never work, because Gyro A was turned off and thus indicated zero roll velocity no matter how the spacecraft was actually moving. After two hours the confused spacecraft gave up and put itself into safe mode again.

Operators on the ground saw from the data sent by SOHO that Gyro A indicated no roll velocity, while Gyro B said it was rotating. Not knowing that Gyro A was not working and basing themselves on previous safe mode experiences, they decided that Gyro B had to be malfunctioning.

Turning Gyro B off, they left SOHO without any working gyroscopic unit. Losing direction, the spacecraft was no longer able to point its main, high-gain antenna to Earth and its solar arrays to the Sun. As a result, control of SOHO was, abruptly and unexpectedly, lost.

Frantic signaling by the powerful antenna dishes of NASA's Deep Space Network did not manage to re-establish contact with the lost satellite. Then the international rescue team, quickly formed after the loss of contact, concocted the idea of using the world's largest antenna, the giant Arecibo radio telescope. This enormous, 305-meter (1,000-foot) diameter dish built into a natural valley on Puerto Rico was used to transmit a powerful signal to SOHO. NASA's large Goldstone antenna was then used to listen for signals returned by the spacecraft, as the basically stationary Arecibo could not be pointed to the spacecraft for very long and only had a transmitter installed at that time.

To the relief of the operators, a faint heartbeat of SOHO was finally received. SOHO signaled that it was slowly spinning and out of battery power because the solar arrays were pointed in the wrong direction. Without power, the heating system had shut down and the thruster's propellant had frozen.

The operators then instructed SOHO to use every bit of available power to slowly thaw out the propulsion subsystem. Unfortunately, every time the batteries were recharging the onboard computer had to be turned off and the software patch containing the new power management instructions was lost from its memory. The software thus had to be re-sent and re-installed every time power levels increased sufficiently to wake up the computer.

After two weeks of fine-tuning, the rescuers finally managed to thaw all the propellant to enable the thrusters to be used for attitude control again. SOHO was then re-pointed in the right direction, and normal operations were finally resumed.

Even though the scientific instruments and other vital onboard equipment had been exposed to temperatures varying from –120 to +100 degrees Celsius (–184 to +212 degrees Fahrenheit) because of the inoperative thermal control subsystem, all except the gyros had survived the ordeal undamaged. The delicate wiring of the gyroscopes had not been able to cope with the extremes and could no longer be used. Fortunately the engineers and operators found a way to stabilize the spacecraft without the gyros, using only the other sensors on board that could still work.

Thanks to the ingenuity of the people involved, SOHO was saved, and is still operating today.

HOPE IS LOST

Sometimes spacecraft are lost despite valiant and daring attempts to save them. Such is the story of the Japanese Nozomi; a bit sad, but also a testimony to the ingenuity of the Japanese space scientists and engineers.

On July 4, 1998, Japan became the third nation ever, after the United States and the Soviet Union, to send a spacecraft to Mars. The probe was called Planet-B, but renamed Nozomi after its launch from Japan's Kagoshima Space Center. Nozomi means "Hope" in Japanese, and hope would soon be needed.

Its M-5 launcher was not powerful enough to speed the spacecraft toward Mars, so Nozomi first went into a long, elliptical Earth orbit that took it around the Moon. Nozomi completed two lunar swingbys to gather speed, then left the Earth–Moon system and entered an orbit around the Sun.

To further increase its velocity, Nozomi was to make a gravity-assist flyby of Earth together with a 7-minute burn of its bipropellant engine on December 20, 1998. Disappointingly, however, a valve malfunctioned and precious propellant was lost, but most importantly Nozomi did not receive a sufficient boost to reach Mars.

In an attempt to get things back on track, controllers at Japan's Institute of Space and Astronautical Sciences (ISAS) activated Nozomi's thrusters for a correction maneuver on December 21. Unfortunately, the burn emptied the tanks so much that not enough propellant was left for the later required braking maneuver near Mars. Without braking, Nozomi would just speed past the planet instead of going into orbit around it.

The situation looked bleak. Nevertheless, the ground controllers found a way to save the mission. They formulated a new flight plan according to which Nozomi would make three trips around the Sun and two additional Earth flybys. These gravity assists from the Sun and Earth would give Nozomi just the right velocity to reach the red planet by 2004, with a low enough speed for it to go into Mars orbit using only a little bit of braking thrust and propellant.

The arrival would be four years late, but ISAS scientists believed that the spacecraft's science instruments would then still work properly. Looking on the bright side, they even saw a benefit of the longer flight: there would be extra time to investigate the solar wind in interplanetary space.

Then, in April 2002, the radio went down due to a massive solar flare that damaged the onboard communications and power subsystems. Such powerful energy outbursts from the Sun send charged particles into space, which can cause radio and electrical disruptions even on Earth.

Fortunately, some contact with the spacecraft was still possible and the probe's computer was not damaged. Engineers were eventually able to fix the problem with the communication subsystem. However, the solar flare had also created an electrical short that had damaged the attitude control heating, which allowed the hydrazine propellant to freeze.

The propellant thawed out when Nozomi got closer to the Sun for an Earth flyby in December 2002, and the thrusters could be used to correct its trajectory. Another Earth flyby in June 2003 also worked well.

However, as it approached Mars five years behind schedule and with nearly empty tanks, the hydrazine froze again due to the increased distance from the Sun. When Nozomi finally reached its destination, it was thus unable to fire its attitude control thrusters and larger main thruster to move into an orbit around Mars. Instead, Nozomi flew on and got stuck in a long, useless orbit around the Sun. It was a sad ending to Nozomi's interesting but tragic life.

NEAR Eros

In February 2000, NASA's Near-Earth Asteroid Rendezvous (NEAR) probe became the first spacecraft ever to go into orbit around an asteroid. NEAR spent a year studying the irregularly shaped, 33-kilometer-long (21-mile-long) asteroid Eros, then its main job was done.

The spacecraft's operators came up with a daring and exciting plan that would result in a scientific bonus and a new space record: landing NEAR, as the first machine ever, on the surface of an asteroid!

The spacecraft had not been designed for this, but the extremely low gravity of Eros would be a great help; rather than a genuine landing, the feat would be more like docking with another, very large, object in space.

The probe was brought down closer and closer to the surface, all the time making ever more detailed pictures. On February 12, 2001, NEAR Shoemaker made a gentle, perfect three-point landing on the tips of two solar panels and an edge of the spacecraft body. The touchdown velocity was only about 6.5 kilometers (4 miles) per hour, low enough to prevent the probe from bouncing back into space. The touchdown was so elegant that NEAR was still operating and communicating with Earth even after landing.

To benefit from this unexpected opportunity for even more bonus science, the operating team got a 10-day mission extension and then a further four days of Deep Space Network ground antenna time to collect all the data sent back to Earth.

The extension allowed NEAR's gamma-ray spectrometer to collect data from its new, favorable position so close to the surface. The instrument's team quickly redesigned the necessary software and uploaded it to the spacecraft. Never before had a gamma-ray experiment operated on the actual surface of another planet. The result was very detailed readings of the composition of the asteroid's surface, enabling a much better comparison with some meteorites found on Earth that are thought to have originated from asteroids.

NEAR Shoemaker now sits silently just south of a saddle-shaped feature named Himeros, a lasting monument to the ingenuity and adaptability of human intellect.

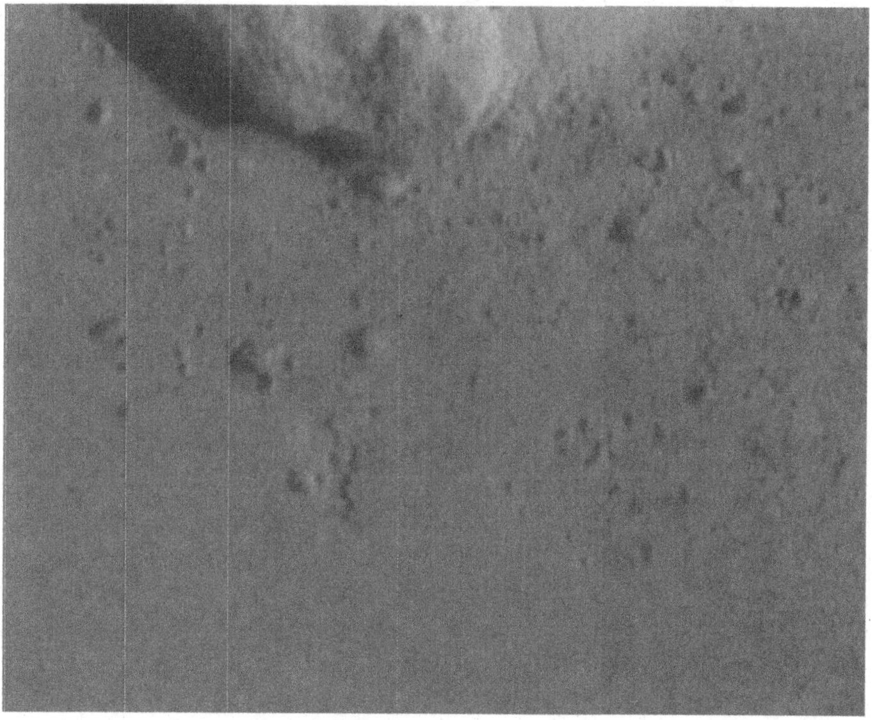

FIGURE 8.2 The last picture made by the NEAR Shoemaker, made from 120 meters (394 feet) above the surface of Eros. The large rock is about 4 meters (13 feet) in diameter. The streaky lines were caused when the actual landing disturbed the transmission of the image. [NASA]

9

A BRIGHT FUTURE

THE opportunities for new scientific discoveries in the Solar System remain boundless. We will continue to send out new, more efficient robotic explorers incorporating exciting new technology with ever-increasing capabilities. We will use them to visit new worlds as well as revisit places we've already had a glimpse of, to see them in new ways and learn much more.

We are going to build spacecraft that can do more science for the same sizes, masses and power levels of our current spacecraft. In parallel, we will develop robots that can do the same amount of exploration with much smaller systems than we are using now. Our mechanical adventurers will investigate regions on planets and moons that were inaccessible for earlier spacecraft, and they will be more adaptive to changing situations.

Some new developments are the result of a "requirement pull"; they are undertaken to reach a certain specified goal. The technology of the Apollo Moon program is a clear example of this: to fulfill president Kennedy's proclamation "of landing a man on the Moon and returning him safely to the Earth," NASA and US industries had to invent new moonsuits, lunar landers, rocket motors, smaller onboard computers and much more. A current requirement pull is the ongoing development of new launcher technology in a quest to make access to space cheaper.

However, more often new space projects are a result of a "technology push"; new inventions making it possible to do robotic space missions faster, better or cheaper.

The fast progress in miniaturization and increased sensitivity of all kinds of electronic detectors has for instance resulted in new scientific instruments that are much better and yet smaller than those launched 10 years ago. And these were in turn already a big improvement over the instruments flown on board spacecraft in the 1980s.

Modern instruments can make more detailed observations than earlier versions, but we need more stable satellites to fully benefit from the improved capabilities. This means the AOCS (Altitude and Orbit Control System) equipment needs to be more accurate. Furthermore, because of the increased amount of detail that the instruments are able to see, they collect larger amounts of data. In turn, this requires improvements in the Data Handling subsystem.

Due to this kind of processes, interplanetary spacecraft on the whole are improved continually; each new mission is a step forward in science as well as in technology.

Nowadays the "pushing" technology often comes from non-space areas. For instance, commercial computer technology has made quantum leaps in the last 20 years or so; spacecraft processors are now based on home PC technology and not the other way around.

The computer currently sitting on your desk is likely to be a generation or two more advanced than those flying on the newest space probes, because it takes time to adapt commercial equipment for use in the radiation and vacuum of space.

Likewise, the development of revolutionary, air-breathing engines for efficient, reusable launchers is more likely to come from the aircraft industry than from the traditional launcher developers.

It is always hard to forecast the future, but nevertheless this chapter attempts to provide a glimpse of what may come. Space agencies, industries, research institutes and universities are constantly seeking out new technologies and inventing innovative spacecraft concepts. There is much more going on than can be captured within the pages of this book.

Some possibilities may become reality soon, some will require long periods of further development, and other dreams may never be fulfilled. However, for every vision of science and technology that fails to materialize, several as yet unforeseen possibilities will arise.

FASTER, BETTER AND CHEAPER?

The trouble with space exploration is that we always want to launch more probes with ever more complex instruments. But as the budgets available are very limited, tough choices have to be made on what missions should be on top of the wish-list and what missions, however interesting, need to be forsaken to free money for the selected few.

Then, at the level of individual projects, more cost-related decisions must be made: Can we afford the scientifically most capable design, or do we have to go for a less preferred but much cheaper option? How much science can we shave off to save cost without dramatically impairing the value of the mission?

Often the version of the space probe that actually gets launched is a lot leaner than the dream-machine presented in the first engineering concept drawings.

In the early 1990s the then head of NASA, Dan Goldin, launched the now infamous "Faster, Better, Cheaper" initiative. According to this edict, NASA would from then on aim to develop spacecraft in less time, with better capabilities, and for lower costs than ever before. There would be strict schedule and cost constraints, a maximum use of already existing equipment and a minimization of management-oversight.

The concept started off well, with very successful missions such as Mars Pathfinder and NEAR Shoemaker being launched shortly after starting the development, and for relatively low budgets. But then a series of disasters followed.

In 1997 the Earth satellite Lewis went spinning out of control soon after launch, due to a flawed attitude control subsystem. Unable to point its solar arrays to the Sun, it could not recharge its batteries and was lost.

One and a half years later the WIRE infrared astronomy satellite also ran into serious trouble shortly after orbit insertion. The cover of its telescope ejected prematurely, making the solid hydrogen that was to act as coolant evaporate rapidly. The venting made WIRE spin out of control, and without proper cooling it was unable to measure the weak heat-radiation of stars it was supposed to detect.

In the same year, Mars Climate Orbiter and Mars Polar Lander were lost within a few months of each other, both due to apparently simple but crucial mistakes that could have been avoided with a bit more testing and project supervision.

Part of the "Faster, Better, Cheaper" plan had been the acceptance of higher risks to bring costs down. NASA had assumed that the higher number of successful missions flown would compensate for the few additional disappointments. However, more spacecraft were lost than expected, and many due to apparently simple and basic mistakes.

The series of failures sent waves of anxiety throughout NASA and the space industry. Various investigation boards concluded that the push for low costs and short schedules had crossed a critical limit. Developing teams were simply not given enough time and money to do their work properly.

A study of The Aerospace Corporation plotted the relative complexity of space missions against the amount of money and time available for their development, and showed that all the recently failed missions were in the "danger zone." The projects had been too complex to be developed within the very limited timeframes and financial budgets allocated by NASA.

As a result, the designing engineers got stressed, the technicians didn't have enough time for all the tests required and the project management lacked the resources to properly check everything that was being done. The final verdict was that the new missions had been done too quickly and too cheaply.

The current consensus in the space industry is that out of the three goals in the "Faster, Better, Cheaper" list, you can only get two at the same time. You can do a mission faster and better, if you are willing to put additional money into it. Alternatively, you can develop your space probe cheaper and faster if you are willing to trade on its capabilities. Or you do it both better and cheaper, but you allow more development time. But as Meat Loaf sings, "Two out of three ain't bad."

However, in a more reasonable form the "Faster, Better, Cheaper" concept still survives. Rather than do only a couple of extremely complex missions per decade, NASA and ESA are now launching many more smaller missions based on greater amounts of standardized equipment.

Nearly every two years, when the planetary alignments are favorable, NASA is sending a probe to Mars. Furthermore, over the last couple of years ESA has been launching a record number of interplanetary spacecraft, and it has been able to do that despite shrinking science mission budgets. It has never been busier in interplanetary space!

GOING COMMERCIAL

In science fiction books, enterprising astronauts and company property robots are already exploring the Solar System on a grand scale, establishing mining bases along the way and capturing asteroids full of precious metals to sell to industries on Earth. In reality, space exploration is still the exclusive domain of government agencies, and private business has remained stuck in Earth orbit.

While commercial companies are operating some Earth observation satellites and most of the world's telecommunications satellites, the vast majority of missions to the Moon and further have been strictly government-funded and government-operated.

This is not very surprising. With a telecom satellite, investors can get a nice profit within a couple of years. However, scientific data from a spacecraft orbiting Jupiter can only be sold to a relatively small group of rather poor scientists and budget-constrained space agencies. And even then, only many years after the initial investment.

Moreover, planetary exploration requires very complex, often unique equipment specially developed for a certain mission. In contrast, most commercial Earth orbiting satellites are based on existing and standard space equipment. A telecommunications satellite can be put in operation within a few years after order, while the development of planetary robot explorers usually takes much longer, requires much more money and involves a lot more technical and financial risk.

No wonder private investors put their money in telecom and Earth observation satellites rather than try to make a meager, late and very uncertain profit by launching risky planetary explorers.

Apart from developing spacecraft under government contracts, there does not seem to be much commercial opportunity in space exploration at the moment. Even if you could manage to find a patch of the Moon covered with bricks of gold, it would not be worth the investment of building and flying a spacecraft to scoop it up and bring it back to Earth. The costs would far outweigh the value of the gold.

Nevertheless, there may be some near-term niche-markets for innovative space missions beyond Earth orbit. For instance, California-based TransOrbital Inc. is a private company that is developing a lunar probe called Trailblazer. Funding and running the project is completely independent of any government agency. Instead, the mission is sponsored by a number of companies that are mostly in it for the advertisement possibilities.

Computer manufacturer Hewlett Packard, for instance, will supply the onboard computer for the Trailblazer spacecraft. It wants to use the project to advertise its new wireless computer technology that will enable Trailblazer's computer to be in contact with all the equipment on board via a radio link. It will also make it possible for anyone with the right hand-held computer to send e-mails to the spacecraft while it is orbiting the Moon, and to get a confirmation that their message has arrived (what the spacecraft will do with the e-mails is not clear, however).

It is even more extraordinary that many private persons also finance the mission by paying money to put all kinds of small personal items on board the spacecraft. Sending a small message to the Moon on Trailblazer costs you $17, launching a business card costs you $2,500, while for anything else you will need to put down $2,500 per gram. There is room for some 10 kilograms (22 pounds) of personal items on board Trailblazer.

Thousands of people have already taken advantage of this unique possibility, making the initiative an important part of the financing of the mission. The whole project is estimated to cost about $20 million, including launch.

Trailblazer weighs only a mere 110 kilograms (240 pounds), which makes it light enough to be launched with a cheap, Russian Dnepr rocket. The Dneprs are former nuclear ballistic missiles that have to be destroyed according to new arms-control treaties. Rather than simply scrapping the original Soviet SS-18 rockets, they are converted to Dnepr space rockets and commercially sold for launching small satellites (nuclear warhead not included, of course).

A smart move: with a launch price of only 8 million dollar they are rather cheap, Russia gets rid of its missiles without having to pay for scrapping them, and its military launch teams can continue to practice launching rockets.

After launch the can-shaped, rotation-stabilized Trailblazer will orbit the Moon for about three months in a very low orbit. Its camera will make high-resolution pictures of the lunar surface that TransOrbital intends to sell to scientific institutes. Additional revenue could come from a new lunar map, based on the Trailblazer images. This will have the highest resolution ever and could help space agencies and private firms to plan future lunar missions.

Apart from stills, Trailblazer will make HDTV-quality videos to be sold for use in documentaries and television commercials.

Trailblazer will also try to image the sites of the Apollo landings. TransOrbital hopes to find the lower parts of the Lunar Modules that stayed behind after the astronauts left, or at least the disturbances of the lunar soil caused by the rocket exhaust of the Apollo landers. There are still many "conspiracy theory" enthusiasts who claim that the Apollo program was an elaborate hoax and that all the moonwalks have been staged in a movie study. Pictures of Apollo spacecraft on the Moon would silence them, and would therefore be very sellable to NASA and to newspapers and magazines.

The gravitational field of the Moon is rather uneven, due to mass concentrations of denser rock, called "mascons," associated with the dark, basalt-filled impact basins such as Mare Imbrium. This will eventually disturb the low orbit of Trailblazer so much that it will fall down and crash on the Moon. All the personal items on board will be packed inside a crash-resistant capsule that is expected to bury itself 4 to 5 meters (15 feet) deep inside the lunar soil upon impact. There the messages, business cards, letters, hairs and other items will remain until found by some future lunar explorers.

Not everyone is happy with the Trailblazer initiative. Some environmentalists are concerned about the Moon becoming polluted. Without an atmosphere and erosion, the metal skeletons of discarded spacecraft and the scars of impacts will remain visible for millions of years. They are objecting to the plan to crash Trailblazer just to put some souvenirs on the Moon.

In reaction, the US State Department has listed some requirements on the mission to minimize the pollution of the Moon. It has also decided that TransOrbital is compliant with these requirements, making Trailblazer the first ever privately organized interplanetary mission approved by the State Department.

Apart from the plans of TransOrbital, it is pretty quiet on the front of private interplanetary missions. In 2000 another company, LunaCorp, got $1 million in backing from the Radio Shack chain of electronics shops. LunaCorp is developing a robotic spacecraft that would be assembled at the International Space Station and from there launched to the Moon. Like Trailblazer, it would generate high-resolution pictures and video and involve public participation. However, the project is still looking for further funding to continue the development.

They also have a plan to develop a small rover that the general public could drive by remote control. Needless to say this is much more complicated than putting a relatively simple satellite in orbit around the Moon.

In the 1990s, Dutch astronaut Wubbo Ockels was leading an ESA project for a Moonlander that would have been partly funded by private sponsors and the general public. However, budget constraints put an end to that innovative plan.

Although some small companies like TransOrbital and LunaCorp may find ways to do interplanetary exploration cheaper and faster than NASA and ESA, as long as spaceflight equipment and especially launches are prohibitively expensive their efforts are likely to remain rather limited. Complicated, expensive projects with long development schedules such as Cassini and Rosetta are likely to remain the exclusive domain of government space agencies for a long time.

FUTURE LAUNCHERS

One reason why interplanetary missions are so expensive is the cost of launching. At the moment the only way to put interplanetary probes in space is by putting them on top of large, conventional rockets. Launch prices for these vehicles are very high, because they can only be used once. The Space Shuttle, even though it is partly reusable, is even more expensive. In general, a launch gobbles up about a quarter to a third of the total financial budget for a robotic space exploration mission.

At some time in the future we may use more efficient, fully reusable rocket vehicles like spaceplanes to put things in orbit at lower costs. Ideally, such a vehicle would not throw away rocket stages on its way up, but fly into space and come back in one piece. Upon return, it should be able be relaunched in a short time, after only a minimum amount of maintenance. Something like a large Space Shuttle Orbiter without the large external propellant tank and rocket boosters, commuting between the Earth and space like a true aerospace plane.

Unfortunately such ideal vehicles are very difficult to build. They need to have extremely light structures to be able to make it into orbit without the benefit of losing weight on the way up by jettisoning empty stages. Moreover, they require very sophisticated engines that are a mix of jet and rocket technology, and advanced thermal protection materials for fully reusable, low-maintenance heatshields.

Most space agencies and major launcher industries are more or less seriously developing such vehicles. In 2004, NASA successfully tested the X-43A, an almost 4-meter-long (13-foot-long) uncrewed, air-breathing vehicle that uses oxygen from the atmosphere to achieve high speeds. On its last flight, in November 2004, it reached a velocity of 9.6 times the speed of sound!

Such velocities are sufficient to compress the air in the engine without the need for any moving mechanisms, as in a regular jet engine. Spaceplanes with such engines, called scramjets, could use atmospheric oxygen during a large part of their flight to orbit, and thus would need to take less oxygen with them than normal rockets. Unfortunately, scramjets only work at hypersonic velocities; at speeds lower than five times the speed of sound (Mach 5) the air does not get compressed sufficiently for efficient combustion and propulsion.

A revolution in space transportation may be coming, but it will take some time before truly cheap access to space with efficient, fully reusable launchers will be available.

However, there are other possibilities for launching items away from

FIGURE 9.1 The X-43A hypersonic plane during testing on the ground. [NASA]

Earth. For instance, you could use a powerful cannon to shoot your spacecraft into space.

During the Second World War Germany developed a super-gun, named "Vergeltung 3" (Vengeance 3) or "V3", to shoot projectiles to England from the coast of France. The V3 consisted of a long barrel with many side chambers containing explosives. Rather than accelerating the projectile with one powerful explosion, the relatively small side chamber charges along the barrel were ignited as the projectile sped past, each time adding a kick to increase the velocity. Fortunately the super-guns were detected by reconnaissance airplanes while still under construction, and were bombed into oblivion before they could be made operational.

In the mid-1960s a Canadian engineer, Gerald Bull, used Navy guns with 40-centimeter (16-inch) diameter barrels to shoot small projectiles to high altitudes under the US Army's High Altitude Research Program, HARP. The "Martlet" projectiles his team developed incorporated solid propellant rocket engines that were ignited after they left the guns, to get them even higher than the gun alone could shoot them. Some 300 shots were fired, and the Martlets attained a maximum altitude of 285 kilometers (177 miles), well into space (although not fast enough to go into orbit).

Many new technologies were developed for HARP, such as sensors, electronics, batteries and even attitude control systems that could withstand the extreme accelerations of the gun launches.

HARP was stopped for political reasons; the Canadian government discontinued its support of the program, and because Bull was Canadian the US Army could not continue to fund his project on its own. Moreover, any weapon covering a larger distance than 96 kilometers (60 miles) was supposed to be under the control of the US Air Force, not the Army.

Disillusioned, Bull "gave in to the Dark Side" and instead offered his expertise to countries such as South Africa, China and Iraq. Probably because of this, he was shot and killed in Brussels in 1990, only few weeks before parts of a super-gun destined for Iraq were intercepted in England. The assassin was never found.

After the first Gulf War that soon followed, UN inspectors found two partly completed super-guns in Iraq. It is very likely that Bull had designed these cannons and was killed to prevent him from continuing his developments; Iraq could have used a finished Bull gun to bombard Israel with projectiles full of toxic gas. Moreover, Bull was apparently helping to improve the accuracy of Iraqi ballistic missiles, which would not have helped to increase his popularity outside Iraq.

Despite its dark past, super-gun technology could be used peacefully to launch spacecraft. The problem is the extremely high accelerations inside the barrel of the gun, equivalent to over 1,000 times the projectile's weight. Normal launchers only expose their cargos to accelerations of about six times their weight, or 6 "*g*," so conventional spacecraft would be completely squashed to a pancake configuration by the extreme acceleration of such a gun launch.

Only very dense systems that basically cannot be compressed any further can survive launch by a cannon. The development of such equipment is not impossible: electronics and cameras used in military missiles can handle hundreds of *g*'s, and the technology developed for HARP could handle over 10,000 *g*. Nevertheless, most of the usual spacecraft equipment currently in use could not withstand such accelerations.

Moreover, to attain the required orbital speed of at least 7.8 kilometers (4.8 miles) per second, a gun launch with conventional explosives would not be sufficient. The projectile would need at least two additional rocket stages to further increase its speed, and even then the useful payload would not be more than about a kilogram. Only very small, compact "pico"-satellites could potentially be launched with an advanced but conventional super-gun system.

However, instead of with explosions, projectiles could also be launched with the use of very strong electromagnetic fields. An electromagnetic launch gun would use powerful magnetic repulsion forces, generated by strong electric currents, to push and accelerate projectiles along a rail. Such systems can potentially achieve higher velocities and launch heavier projectiles than traditional gun technology.

Apart from launching small rockets and satellites, electromagnetic rail guns could assist the launch of larger, more conventional rocket vehicles. A spaceplane accelerated on a electromagnetically powered rail system would get a modest but useful initial kick, and thus require less propellant to get into orbit.

The downside is that even a small electromagnetic launch system requires very high electrical power levels and very long barrels or rails. Moreover, it means that the vehicle will already have a very high velocity at low altitude, where the atmosphere is very dense and the aerodynamic drag is thus high.

Potentially useful for launching much heavier spacecraft are nuclear fission rockets. At the end of the 1950s American scientists working on nuclear weapons developed a method of launching a giant spaceship by detonating a series of atomic bombs under it. This "Orion" spaceship was supposed to fly astronauts to the Moon, Mars and even the moons of Jupiter and Saturn in the late 1960s and early 1970s.

With intervals of less than a second, Orion would shoot out hundreds of small nuclear bombs at the back. These bombs would then explode some distance behind the ship, propelling Orion by hitting a special bumper shield, the so-called "pusher plate," with bomb debris. Giant shock absorbers would connect the pusher plate to the rest of the ship to smooth out the pulses of thrust generated by the explosions.

After initial interest by the Air Force and NASA, the project soon lost support because of the huge leap in technology that would be required. Moreover, the public and politicians became increasingly sensitive to exploding nuclear weapons in the atmosphere; launching Orions from the Earth's surface would have resulted in serious amounts of deadly radioactive fallout.

The designers came up with an alternative plan to launch smaller Orions with Saturn V rockets into Earth orbit and have them start their trip to Mars or beyond from there. However, at the time NASA was already working on another nuclear propulsion concept called NERVA, and thought Orion was a step beyond that, for later in the future. The only other potential customer, the US Air Force, liked the idea of a giant nuclear spaceship but could not suggest a credible military role for its use.

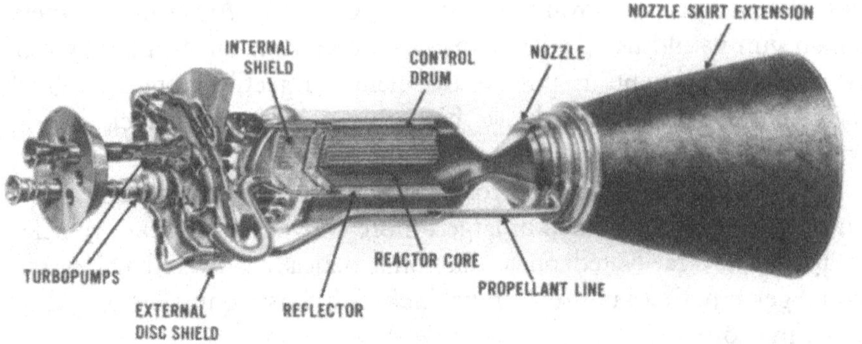

FIGURE 9.2 An overview of the NERVA nuclear engine, which was developed in the 1960's and 1970's. [NASA]

Time caught up with Orion and the project was abandoned. The concept may be resurrected at some point in the future, but certainly only for use far away from Earth.

NERVA was a more modest and realistic US nuclear fission propulsion project. During the 1960s and 1970s experiments were performed on NERVA nuclear fission rockets in the United States, primarily for use in crewed spacecraft bound for Mars.

Energy from a nuclear generator was used to heat hydrogen gas to extremely high temperatures. The hydrogen was then expelled through a conventional rocket nozzle. NERVA's concept was proven to work and the technology is quite mature, but the work has been all but abandoned. Soon after the last Apollo Moon missions, it was decided that no astronauts would be sent to Mars in the near future, so NERVA's power was no longer needed.

Anyway, it is very unlikely that nuclear fission rockets will ever be allowed to operate within the Earth's atmosphere because of the radiation pollution risks involved. A nuclear space explorer would probably have its reactor launched in small elements; the critical mass of radioactive material needed to start a nuclear reaction would then only be achieved when all the pieces are put together in space.

Nuclear fission is based on processes that generate energy when atoms break apart. Instead, nuclear fusion is based on energy radiated by the merging of atoms. The Sun works on this principle. Fusion processes generate much less harmful radiation and radioactive material than nuclear fission, and fusion rockets could thus be a lot safer.

However, fusion technology development has only just started. No one has yet even succeeded in building a nuclear fusion reactor that provides more energy than has to be put in to sustain the fusion reaction. The time when such reactors are mature enough for use as spacecraft propulsion systems, is surely quite some years away.

Another very interesting technology that has just entered the experimental stage is laser propulsion. The idea is to use a ground-based laser to heat propellant on board a vehicle, with the resulting superhot gasses being expelled through a rocket nozzle. Because of the potentially higher exhaust velocity, the vehicle could require much less propellant to reach orbit, while all the energy-generating equipment could remain on Earth.

The downside is that the laser used has to be very powerful, even for launching small satellites into orbit. The development of this launch concept is therefore limited by the power of the lasers we can build today, and this is by far not sufficient to put anything into space.

In the USA, experiments are performed with small projectiles that focus incoming laser light on the air under them. The laser basically makes the air explode and this boosts the metal object upward. Altitudes of several tens of meters have been achieved so far – a promising start but only a baby step in comparison to what is needed to launch a satellite or interplanetary space probe.

An even more exotic way to reach orbit is the space elevator concept. According to a NASA study, a cable of an incredibly strong material could be used to connect an orbital space station directly with the Earth's surface. The center of mass of the system would have to be positioned in Geostationary Earth Orbit (GEO), where its orbital rotation is 24 hours and it stays over the same point above the equator as the Earth rotates around its axis. In any other orbit the system would move too quickly or too slowly and break the cable.

Electromagnetically suspended elevators or mechanical climbers could shuttle between the surface and stations along the cable, without the need for any rocket propulsion. Once transported to the main hub, at GEO altitude, satellites and space probes could be launched directly from the space elevator. As these objects would start with the same velocity that the space elevator is traveling at that altitude – which is GEO orbital velocity – the rocket systems required to send spacecraft to the planets would thus be very small.

The feasibility of the space elevator concept depends mostly on the development of materials that can withstand the incredible forces on the cable. Steel is too heavy and by far not strong enough. Recently developed nanotube materials, based on cylindrical molecules of carbon, may make it feasible. However, the longest nanotube cables made so far are just about a meter long, so we are not yet able to produce the very long cables that would be required.

Once a space probe has already been placed in low Earth orbit by another means of transportation, a system of freely orbiting and rotating cables or "tethers" could be used to transfer spacecraft to higher orbits. Such a system would consist of a very long tether, 100 kilometers (62 miles) or more in length, with a ballast mass at one end and a sort of spacecraft catching or docking device on the other.

As the cable would be in orbit around the Earth, it would be spinning like a sling, with its center of rotation near the ballast mass. A spacecraft in a lower orbit could hook itself onto the docking end of the tether and let itself be swung into a higher orbit. With a series of these tethers, or "sky hooks," spacecraft starting just above the atmosphere could be shot all the way into an interplanetary orbit.

A spacecraft accelerated in this way does so by stealing momentum from the tether. By transferring energy to the spacecraft, the tether's own orbit is lowered. Eventually, it would fall back down to Earth if nothing was done to reboost the system.

Restoring the tether's orbit could be achieved by lining the long cable with metal wire and passing an electric current through it. The electric wire would generate its own magnetic field, and the interaction of the wire's magnetic field with the Earth's magnetic field would result in a small but steady force that could push the tether back up.

The power required could be generated by solar panels. This concept of generating a force with a cable and a current has been tested during a Space Shuttle mission, and was found to actually work quite well (except that the cable burned through due to an electrical spark).

SOLAR SAILING

Getting space probes out of Earth orbit and accelerating them to high speeds to reach distant planets in a reasonable time costs a lot of propellant. This means big, expensive rockets and large and heavy propellant tanks on the spacecraft. To limit launch costs and leave enough room for an interesting suite of instruments, most modern interplanetary space probes use slingshot planetary flybys to increase velocity and change orbit, but that takes a lot of time and can involve risky maneuvers.

There is, however, another way that may completely eliminate the need for rocket propulsion on board a space probe.

Some 400 years ago, the famous astronomer Johannes Kepler noted that the tails of comets are swept away by some mysterious breeze from the Sun. He imagined that it might be possible to use this to propel spaceships, the way sailing ships on the ocean are moved by the wind.

Although Kepler's version of solar wind doesn't really exist, we now know that light exerts a tiny force on anything it is reflected by. The push is extremely small, less than 1 kilogram force per square kilometer (0.8 pound force per square mile). On normal spacecraft it can thus be completely neglected, but with a big enough surface it becomes significant.

A large, ultralight spacecraft with enormous, mirror-like solar sails made of extremely thin foil would be able to be pushed around the Solar System by Sunlight. Slightly similar to electric propulsion, the push would be very, very small but it would act continuously, as long as the Sun was close enough and able to shine on the sail. Eventually, the continuous force of the light on a solar sail could propel a spacecraft to speeds several times faster than is possible with traditional rockets.

The continuous force of Sunlight acting on a solar sail could also be used to put spacecraft into very special orbits that are no longer strictly limited to the orbital mechanics rules of Newton or Kepler. A solar sail may, for instance, hover in a fixed position with respect to the Sun, balancing gravity and light pressure forces against each other. Computer simulations show, however, that the accurate control of such a solar sail would be very tricky.

The use of large lasers on Earth to help to push the spacecraft could enhance the effectiveness of solar sails, but very powerful lasers would be required. Moreover, a laser loses its usefulness over long distances because the beam diverges.

The major problem with solar sailing is how to fold the vast surface of ultra-thin foil in such a way that it fits inside the payload fairing of a launcher, and can still be deployed without damage once in space. Both NASA and ESA have made some deployment tests of small solar sails on Earth, and are studying demonstration missions to test deployment in Earth orbit.

The Planetary Society, an organization of space enthusiasts in the USA, has already attempted to launch such a test model. Its fully privately financed Cosmos 1 spacecraft was designed and built together with the Russian Lavochkin Association and the Russian Space Research Institute.

Looking a bit like a windmill, Cosmos 1 had eight triangular sails of 15 meters (50 feet) in length. The spacecraft was launched with a Volna rocket – a converted ballistic missile left over from the old Soviet arsenal –

FIGURE 9.3 The Cosmos 1 solar sail as it should have looked in Earth orbit. [Planetary Society]

from a submerged Russian submarine in the Barents Sea on June 21, 2005. Unfortunately, the engine of the first stage failed less than 2 minutes into the flight and Cosmos 1 fell back to Earth.

If it had made it into orbit, the solar sail would have been deployed by inflatable tubes. To prove the principle of solar sailing, the satellite would then have attempted to gradually raise its orbit by sunlight pressure alone. The Planetary Society had hoped that their demonstration would have opened the way to interplanetary space travel by solar sail, and would some day have led to a probe sailing to the stars.

The inflatable tubes used on Cosmos 1 are also a novel space technology. Inflatable structures can be squeezed into small canisters and then inflated in space using tanks with compressed air, or gas generators that make gas through chemical reactions. Apart from use in long booms for solar sails, inflatables could be employed for huge dish-antenna space radiotelescopes and other large structures.

The inflatable material can be configured so that it eventually hardens, for instance due to the ultraviolet radiation from the Sun. That ensures that the structure remains rigid even if it loses pressure owing to tiny punctures by micrometeorites.

Apart from the Cosmos 1 booms, there has been other experience with inflatable technology. In 1996, the Space Shuttle Endeavor successfully deployed a 14-meter (46-foot) prototype inflatable antenna in Earth orbit. In 2000 ESA and Russia launched their Inflatable Re-entry and Descent Technology (IRDT) demonstrator – a capsule with an inflatable re-entry heatshield. Initially it worked quite well, but unfortunately at lower altitudes a tear occurred in the inflatable shield, resulting in a higher velocity and a heavier than expected impact on landing. The IRDT capsule was damaged, but the valuable test data in its computer memory could still be retrieved.

Hopefully, demonstrators such as Cosmos 1 and the ongoing development of solar sail and inflatable technology will make it possible to launch real, operational solar sail probes sometime in the not too distant future. It is an elegant way of exploring space, reminiscent of the sailing ship expeditions of discovery in our past.

Nuclear Energy

Going out in the Solar System, beyond Mars, getting enough power is always a problem for probes equipped with solar arrays. That is why such spacecraft as Cassini and Galileo have RTGs that use the heat given off by slowly decaying radioactive metals to obtain electricity.

However, a true nuclear reactor with a sizzling fission process would give much more power. Where spacecraft usually have to contend with a couple of thousand watts at most, a nuclear space reactor could supply tens to hundreds of kilowatts. Such large amounts of energy could, for instance, be used to employ powerful radars able to penetrate the thick ice shells of Jupiter's moons.

Relatively small nuclear reactors have already been flown on board Earth-orbiting satellites. The USA has only launched one and that was in 1965; but in the 1960s and 1970s the Soviet Union powered many of its spy satellites with fission reactors.

Until 1993, the Defense Department, the Department of Energy and NASA worked together on a 100-kilowatt reactor called SP-100. It was nearly fully developed and ground tests had even been conducted, but the SP-100 never flew.

Recently NASA has proclaimed renewed interest in the technology for application in interplanetary spacecraft; the Jupiter Icy Moons Orbiter (JIMO) project would have used a nuclear reactor. The fission process was to provide energy for a series of ion thrusters to propel the spacecraft, and would generate unprecedented amounts of power for the rest of the onboard equipment. Powerful new types of scientific instruments, such as large radar and laser systems, could have been used. According to the plans, the spacecraft's power subsystem would give the craft over a hundred times more power than a traditional non-fission system of comparable weight.

However, the project was going to be too expensive and, moreover, it appeared that the project was more driven by the (military) desire to develop a nuclear space reactor than to further explore the Jovian system. Not surprisingly, it was cancelled (although the development of a space reactor continues, because of its uses for powering crewed lunar bases and interplanetary expeditions).

There are many other, more modest and much cheaper ways to explore Jupiter and its moons than by using giant nuclear spacecruisers. Nevertheless, the use of nuclear reactors would make human missions to Mars a lot more feasible. Using them for propulsion would save a lot of propellant, and reactors could be used on Mars to provide energy for long-duration stays.

WORKING TOGETHER

Another very interesting innovation is the idea of splitting the necessary functions over several spacecraft instead of one. ESA has, for instance, been studying a concept for a Jupiter mission called the Jovian Minisat Explorer (JME). At present, JME is a feasibility study, set up to investigate the requirements and constraints that such a future mission may face. Nevertheless, it shows how, in the future, multiple spacecraft may work together.

The JME mission concept involves two small spacecraft that, during launch and transfer to Jupiter, will be attached to each other. Once they reach their destination, they will split up. One will be put into a low orbit around the icy moon Europa, while the other will be flying high above Jupiter itself.

From an altitude of only 200 kilometers (125 miles), the Europa Orbiter will make detailed investigations of the moon and maybe even eject a compact microprobe to study the ice crust on the surface. At the same time, the other orbiter can begin its investigation of Jupiter.

However, the Europa Orbiter will have to work very quickly, because

Jupiter's strong magnetic field is full of charged particles. This results in intense radiation in a wide zone around the giant planet, and Europa orbits smack in the middle of it. The damaging radiation will be a serious hazard for the spacecraft, especially to its sensitive electronics that will have to be particularly rugged and shielded by a thick layer of metal.

Moreover, Jupiter's uneven and very strong gravitational field will continually try to push and pull the probe out of its orbit. With the limited amount of propellant it can carry, the satellite will only be able to fight back for about two months. After that it will lack the thrust to compensate the perturbations and is doomed to crash onto Europa's icy surface.

Because Europa is a place where life may possibly have evolved (under the surface ice), ESA will have to prevent contamination with microbes from Earth. This means that the spacecraft must be subjected to rigorous Planetary Protection sterilization measures, although the severe radiation around Jupiter may help to diminish the number of microbes on board.

Before the crash, the Europa Orbiter will send all the data it has collected via high-data rate radio link to the other spacecraft, which will be in a much safer orbit. This Jupiter Relay Spacecraft will store all scientific treasure on its computer memory, and leisurely send it back to Earth. It can take its time doing that, using low data rates requiring limited transmitting power, because it will be in a much more stable, relatively low-radiation orbit.

GOING FOR A WALK

Mobile rovers greatly enhance the possibilities of a mission. Normally we don't exactly know what we are going to find at a new landing site (otherwise there wouldn't be much benefit from going there in the first place), or where exactly the lander will touch down. Nothing is more frustrating than to land in a scientifically exciting area, only to find that all the interesting rocks are just out of reach of the fixed spacecraft's robotic arm.

With a mobile system, rather than having to land a spacecraft at exactly the right spot, it suffices to land somewhere in the vicinity of an interesting area. The rover can then take the scientific instruments to the preferred location.

Having a rover that is independent from its lander allows the engineers to separate the scientific and mobility functions from those only needed for descent and landing, which are left on the lander itself (and often the functions for flying through space stay behind as well, on a separate part of the spacecraft in orbit around the planet).

Most current robotic planetary surface explorers, such as Lunokhods 1 and 2, the Pathfinder Sojourner rover and the recent MERs, have wheels. The only exception was the "hopper" lander on the Russian Phobos 2, which would jump over the low-gravity surface of the Phobos moon of Mars by rotating two sets of boom "legs." However, the Phobos 2 spacecraft was lost before this small lander could even be detached.

Wheels allow a rather simple mobility and steering system, and you can achieve quite high speeds with them. However, wheels operate best on fairly level, smooth terrain – the kind you generally don't find on any planet except Earth. Caterpillar tracks like those on tanks make it easier to move in uneven terrain, but they result in rather heavy vehicles that require a lot of power.

Nature has never invented the wheel, probably due to the problem of connecting the skin, blood vessels, etc., of a body with rotating and therefore partly separated elements. Nevertheless, in rough terrain nature's solutions for land mobility are all far superior to wheels.

Animals walk, hop, climb, slide and slither over, under and between rocks. They are able to cope with terrains in which our wheeled vehicles get stuck. Just try to get a machine with wheels to climb stairs! And a walking rover would have little trouble getting out of the soft, powdery sand dune that obstructed the Marsrover Opportunity for quite some time.

It therefore makes sense for us to develop walking rovers that, for instance, can explore the Martian surface faster and more extensively. However, to make large robots walk on legs has proven to be very complicated.

If you want your walking rover to always be in a stable position, you need at least four legs and cannot lift more than one at a time. That means a relatively slow vehicle. You can test this yourself: try to walk in such a way that you remain completely stable on one leg up to the moment that your other foot touches the ground. Only then do you move your weight and repeat the process with your other leg. Every step will take you much more time than usual.

When you walk normally, you constantly shift your weight forward and are therefore continually out of balance. If you suddenly stopped during a normal step, you'd fall forward. While walking, you are what we call dynamically stable (i.e. you don't stumble), but statically unstable (i.e. you cannot suddenly "freeze" in a walking position without falling over).

Small, multiple-legged animals like insects typically walk statically stable. At no point during their movements are they out of balance. Because they are very small and light they can still attain relatively high speeds in comparison to their size. However, somewhat larger and faster animals generally have no more than four legs and move statically unbalanced; they run dynamically stable. You would expect that this requires very complicated "software" and accurate measurements of terrain features, but even small-brained animals are very good at it. Somehow they (and we) are able to walk quickly without constantly observing and measuring the terrain on which they walk.

The key appears to be a combination of sensors that recognize "up" and "down" and indicate accelerations (the vestibular system in your head), as well as sensors in the legs that signal their position and when they touch the ground. Instead of depending on detailed maps indicating objects, heights and slope angles, animals rely on real-time reactions to impulses from their senses; they feel their way around, rather than calculate and plan.

Walking robots are a hot topic in robotic development on Earth at the moment. Honda's ASIMO, for instance, is a human-like, human-sized robot that is able to walk on two legs. It can even climb stairs! ASIMO has been specially developed to operate in a human world with houses, doors, tables and light-switches. Furthermore, it can interact with humans; ASIMO can greet approaching people, move in directions they indicate, and even recognize their faces and address them by name. One day a descendent of ASIMO may accompany astronauts on their exploration of Mars, or even be the first to test whether humans can safely walk on a moon of Jupiter.

An example of a simpler system is "Genghis," a six-legged, 1-kilogram (2-pound) walking robot developed at the MIT Artificial Intelligence Laboratory. Each of its legs has two electro-motors that also give it some rough force measurement; if a leg hits anything, the increase in stress in the motors is sensed by the robot. That way Genghis can tell when a leg has touched the ground or hit an obstacle, so that it can keep its balance and is able to navigate over uneven terrain. Two whiskers in front prevent head-on collisions with obstacles in its path.

The Scorpion robot of the University of Bremen in Germany has eight legs, with which it can move through sand, climb steep ramps, overcome obstacles and even walk along a beam, hanging upside down. In one experiment the Scorpion grabbed a plank with one leg, and walked away with it using the remaining seven legs.

If moving relatively fast, walking robotic rovers would need to be much more autonomous than the current wheeled vehicles. If they had always to wait for an operator on Earth to steer them around, their speed limit would be very low. Ideally, an operator should be able to just indicate where he or she wants the rover to go, and the robot should then be able to figure out how.

A first application of this was tested with the Pathfinder rover, and was later also applied to its more sophisticated descendents Spirit and Opportunity. For these rovers, an operator chooses a suitable path on the images sent back by the robots. The operator then indicates a set of target locations along the route, and instructs the rovers to move from target to target, like golfers on a golf course. Additional commands, such as supplementary, in-between target locations or simple instructions to go a couple of centimeters back, only need to be sent when the rovers get blocked by unexpected obstacles.

Even if the next generation of rovers still uses wheels, advanced onboard intelligence, as pioneered with the ASIMO robot, is still a very interesting technology to make rovers explore a planet's surface faster and more independently than they are doing now.

FLYING

Rather than driving or walking, some future robot explorers may be flying over Venus, Mars or Titan. Robotic planes can give us high-resolution, bird's-eye views of the terrain beneath. Moreover, they can investigate the atmosphere like no orbital spacecraft (too high) or surface rover (too low) ever could.

In the 1990s NASA had plans for a small robotic plane to fly over Vallis Marineris, the famous "Grand Canyon" of Mars. It would have inflatable wings, a hydrazine-powered motor and be deployed from its clam-like descending capsule in mid-air. Because the density of the air on Mars is much lower than on Earth for a given altitude, the plane had to be designed with relatively large wings and an oversized-propeller.

The vehicle was meant to reach Mars in 2003, to commemorate the first powered flight of an airplane by the Wright brothers a hundred years earlier. Unfortunately the project proved to be too complex and costly, and was cancelled.

Nevertheless, the idea lives on. In 2002, an experimental Mars-plane called "Eagle" was successfully flown on Earth. To simulate the low density and pressure of the air on Mars, the plane was taken to an altitude of 30 kilometers (19 miles) by a stratospheric balloon. There it was

FIGURE 9.4 ARES, a NASA concept for a small plane that is to explore Mars from the air. [NASA]

dropped, unfolded its wings and tail and subsequently made a 90 minute, pre-programmed but unpowered flight back down. NASA hopes such tests will enable it to launch a real Mars-plane sometime after 2010.

The atmosphere of Venus is much denser than that of Mars, making it easier for a plane to stay aloft. At an altitude of 64 kilometers (40 miles) above Venus, the air density is the same as on Earth at a height of 16 kilometers (10 miles). A robotic plane could study the lower atmosphere of Venus from such an altitude, while staying above the high-temperature zones.

Because of the Sun's proximity, a Venus plane could be powered by solar energy rather than a chemical engine. It would also benefit from the very slow rotation and therefore long day of Venus; staying around the same location, the solar cells on the wings could receive sunlight for 50 Earth-days without interruption. In comparison to a Mars-plane operating in a very thin atmosphere, it could also fly with less power.

Balloons can provide another means of flying. They can be used to make measurements of the atmosphere of Venus, for instance, just like the balloons of the Russian Vega probes in 1985. Future balloons may use the temperature differences between the Mars day and night to make measurements of the planet's surface.

During the day the Sun would heat up and thus expand the gas in the balloon. With a resulting lower density than the surrounding air, the system would float up. At night the balloon would cool down, shrink and slowly descend. A small experiment package hanging under the balloon would then touch the surface and make scientific measurements. The next day the balloon would automatically go up again and, driven by the wind, float to a new location. In this way a wide variety of locations could be investigated, while the precise tracking of the balloon would give us information about the wind strength and direction.

Such balloons would be relatively simple and require little power, but would also be difficult to control and steer. Just as on Earth, zeppelins would be much more controllable. Aerodynamically shaped zeppelin-like aerobots with propellers could be used to actively navigate their way around. Covered with solar cells to power electric motors, they could also operate much longer than airplanes that require lots of power to stay aloft (however, these are smaller and fly faster).

Taking advantage of advances in micro-machinery and the relatively low gravity on Mars and Titan, one day we may even send insect-like flying robots to scout these worlds for us. Flapping wings give insects the ability to hover like a helicopter, softly touch down, quickly lift off straight up, and rapidly alter direction. With artificial muscles rather than conventional electric motors or engines, such robots could run on very little power.

MICROBOTS

Apart from the development of low cost, perhaps even reusable launchers, it will also pay to design smaller and lighter spacecraft to bring mission costs down.

"The lighter, the better" applies to nearly everything in spaceflight. The lower the mass of a space probe the less expensive it is to launch, the more velocity it can gain from a rocket and the easier it is to land (smaller airbags, smaller parachutes, less propellant needed for retro-rockets, etc.).

NASA's 1997 Sojourner Marsrover has already proven that with modern micro-electronics and miniaturized spacecraft equipment, even a small spacecraft can perform valuable science. Current designs for Russian microchods and NASA nanorovers are the size of a matchbox and have pushed the minimum volumes required for planetary rovers further down.

The new field of micro-mechatronics will make it possible to shrink all kinds of space systems even more, and therefore drastically decrease the

FIGURE 9.5 Future robotic explorers may be very small, as shown by this NASA nanorover prototype. [NASA]

total mass and volume of space probes. Acceleration-measuring devices of only one square centimeter and gearboxes that can only be seen with a microscope have already been built. Micromachines like these are extremely small, light, cheap and they use very little power.

The use of such technology in spacecraft may have dramatic consequences: a lower spacecraft mass means that you need less propellant for attitude control and propulsion, therefore smaller tanks, resulting in a lower volume and lighter structure that also needs less thermal protection coverings. Snowball effects like these usually make the mass of a conceptual space probe spiral out of control by seemingly minor alterations in the design. However, the application of micromachinery may enable us to benefit from the high interdependability of spacecraft equipment; rather than spiraling outward, it may decrease spacecraft mass figures exponentially.

The resulting microspacecraft could be the ultimate solution to the "Faster, Better, Cheaper" ideal. Such space probes could open a whole new world for space exploration. Instead of building one large, expensive spacecraft, it may be possible to build many more smaller craft for the same price. This would increase the chance of mission success, because the failure of one spacecraft would not mean the failure of the entire mission.

For the future we may foresee swarms of small spacecraft investigating planets. Some may venture in highly dangerous areas, relaying data to motherships orbiting outside the danger zone. If we have a lot of them, some probes may be sacrificed on kamikaze missions to scout the way for other spacecraft.

Like an army of ants, each of the probes in such a swarm could have its own task. Some may be measuring radiation, others may take pictures and again others may be only used for transmitting data between the probes and Earth.

On planetary surfaces, some may have wheels to go far and fast, others may have legs to explore rough terrain or climb steep hills. They could even be combined: a wheeled rover that can move quickly to interesting sites, and deploy one or more legged robots for detailed investigations. A flying robot could be used to provide information on the precise locations of each individual explorer.

Robots with the same hardware may still differ in their operational software, and could react differently to similar situations. While the programming of some probes may lead them into disaster, others will decide to do something different and survive.

As with insects, the efforts or loss of an individual would be of little consequence, but together they may be able to do highly complex missions resulting in loads of new scientific data.

NASA's Institute for Advanced Concepts is studying a system of using microbots for planetary exploration. It would involve the distribution of hundreds or even thousands of centimeter-scale robots over a planet's surface by an orbital mothership. The little probes would hop and bounce around, powered by polymer muscle actuators similar to those envisioned to power insect-like flying robot explorers.

A suite of miniaturized imagers, spectrometers, sampling devices and chemical detection sensors would allow the microbots to investigate their surroundings. Collectively, the micro-explorers would be able to analyze a large portion of a planet's surface. However, significant advances would also have to be made in data processing and analyses techniques, otherwise it may be difficult to turn the enormous volumes of data collected by a legion of microbots into useful scientific information.

Another interesting concept is to inject a large number of small probes into the atmosphere of a planet, to see where the winds at different locations would take them. This could lead to complex three-dimensional maps of wind patterns and atmospheric properties.

In the more distant future, even smaller nanomachines can be envisioned. Nanotechnology involves extremely tiny machines, no larger than a few nanometers, that are built up molecule by molecule. A nanometer is just a billionth of a meter; you'll need powerful microscopes to be able to see them. If we learn how to arrange atoms and molecules one by one, we could create very, very tiny machines with revolutionary abilities such as self-healing.

The bodies of plants, animals and people have the amazing but crucial ability to repair themselves. If your body could not mend a small wound after you cut yourself, you would not survive very long.

Machines on Earth generally cannot fix themselves, but we can do that for them. Not so, however, with space probes exploring the depths of the Solar System. As we cannot reach them, repair is impossible. When something fails, we can only switch to backup equipment, and if that also fails then the mission is over. Wouldn't it be great if we could give our spacecraft the ability to heal themselves? Nanotechnology may make it possible.

Large, fatal cracks in a spacecraft's structure usually start as microcracks that slowly grow due to repeated bending or thermal expansion. One of the first steps in self-repair could be to embed glue-containing microcapsules into spacecraft materials. When microcracks formed in the structure, these would also break the capsules and release their healing glue. The cracks would be automatically fixed before they could grow longer and cause real damage.

But real self-repair would require the equivalent of cells in living beings; nanomachines acting like tiny factories that can produce sophisticated structures from raw materials available around them. They could be released at damaged places in the spacecraft and actively fix damage by repairing the material molecule by molecule. In biotechnology something like this already exists in the form of machines that can create specific sequences of amino acids on demand.

Even more exotic (we are really getting into the science fiction area here) is the concept of the assembler/replicator. These would be machines that are able to make copies of themselves, just as a living cell does. Each copy could then make a copy of itself in turn, so the number of assembler/replicators would increase exponentially. We could then for instance "seed" an asteroid with some of these machines, after which they would start to replicate and "grow" large space structures such as antenna arrays for radio astronomy. Such possibilities appear rather far-fetched at this moment, but just like micro-electronics, nanotechnology may suddenly shift into top gear and cause a revolution on Earth as well as for space exploration.

BEYOND THE SOLAR SYSTEM

Pioneers 10 and 11, and the two Voyager spacecraft, are currently our most distant space probes. Voyager 1 has recently even reached the zone where the solar winds hits the interstellar gas, 14 billion kilometers (8.7 billion miles) from Earth. However, although these four probes sent us valuable data about the Solar System's far frontier, they have not been specifically designed to explore the area they are now flying through.

Scientists would like to make detailed measurements of the heliopause, the blurry boundary between the Sun's immense magnetic bubble (called the heliosphere) and the interstellar gas. What kind of complex interactions are happening in this shock region, where the solar wind crashes into the interstellar medium? Furthermore, they are very interested in what lies beyond this boundary. What kind of particles, sent out by other stars, will we find there?

Both NASA and ESA are doing studies on an Interstellar Probe, possibly using solar sails to reach the heliopause within a couple of decades. However, the spacecraft concepts are still in a very early stage.

Over the past 10 years, astronomers have discovered over 150 so-called exoplanets near other stars. Orbiting planets make a star "wobble," which can be measured by detecting the Doppler effect in the light of the star. Another way to find them is to accurately measure the amount of light from a star; if a large planet goes in front of it, the light will slightly dim.

Most of the exoplanets discovered until now are very large and usually have a couple of times the mass of Jupiter. However, astronomers recently found a relatively small planet with a mass of 7.5 times that of the Earth and twice its radius around the star Gliese 876.

The planet circles its mother star in a mere two days, and orbits so close to it that on its day side the temperature probably soars to around 200 to 400 degrees Celsius (400 to 750 degrees Fahrenheit) – much too hot for life as we know it. The planet thus also seems to orbit too close to Gliese 876 for liquid water to exist, but its atmosphere might contain a layer of dense steam.

Its mass makes it probably too light to retain a very thick layer of gas around itself, so it perhaps more resembles the rocky planets like Earth and Mars, with an iron core and silicate mantle, than the giant gas planets such as Jupiter and Saturn. To date, it seems to be the most Earth-like planet ever found orbiting a star that is not much different from our Sun. Three other supposedly rocky planets have been discovered outside our Solar System, but they orbit a pulsar, the remains of an exploded star that is flashing at radio frequencies.

The discovery suggests that smaller, even more Earth-like planets probably exists too. However, due to their lower mass, Earth-size planets are hard to find with our current telescope technology. For the next decade, NASA and ESA are planning large space telescopes that should be able to find small planets orbiting other stars. Using spectroscopy, and based on what we have learned about the planets in our own Solar System, we may even be able to detect what kind of atmospheres these "exoplanets" have. Could there be planets out there with oxygen-rich atmospheres like Earth?

What about sending a probe, a true interstellar explorer, to another planetary system? It can be done; even the two Pioneers and the Voyager twins are likely to reach the vicinity of other stars on some very distant day. However, to bridge the interstellar void within a human lifetime is extremely challenging.

Let's return to the miniature Solar System introduced in Chapter 7. With a grapefruit acting as the Sun, Earth is a grain of sand orbiting 10.5 meters (34 feet) away. Mars orbit would be at a distance of 16 meters (52 feet), and the outermost planet Pluto would circle at an average of 414 meters (1,359 feet) from the Sun.

The closest star, Proxima Centauri, would then need to be placed at a distance of about 2,900 kilometers (1,800 miles). With the Sun being a grapefruit located in New York, our neighboring star would be found somewhere near Denver. Locating the Sun in Dublin would put Proxima Centauri near Moscow.

The distance to Proxima Centauri is 4 light-years and 4 light-months; meaning it takes light 4 years and 4 months to get there. Voyager 1 is currently traveling at 17.5 kilometers (11 miles) per second, meaning that it could cover the 150-million-kilometer (93-million-mile) distance between the Earth and the Sun more than 3.5 times in a year. If we were to launch a probe straight at Proxima Centauri with an average speed similar to Voyager 1's current velocity, it would take about 750 centuries to arrive! That is a bit long to wait for scientific return.

There are ideas for extremely fast interstellar spaceships, all based on rather hypothetical means of propulsion that we will not be able to build for many years. A concept based on technology that might be available sometime this century is that of the Daedalus uncrewed interstellar space probe.

Daedalus is the result of a study performed between 1973 and 1978 by a small group of scientists and engineers belonging to the British Interplanetary Society. They started with the requirements that the spacecraft had to use current or near-future technology, that it would be able to reach its destination within 50 years, and that the design be flexible enough to be sent to any of a number of target stars.

The selected baseline target was Barnard's Star, a red dwarf star 5.9 light-years away. It was chosen over closer Proxima Centauri because astronomical data at the time suggested that Barnard's Star might be orbited by at least one planet (but this data was later found to be unreliable).

To reach Barnard's Star in 50 years, the spacecraft would need to cruise at about 12 percent of the speed of light. Such a velocity cannot be achieved by conventional rockets, so the team had to look for a more advanced, and therefore not yet existing, means of propulsion.

They chose to power the starship by nuclear fusion. Small pellets of deuterium and helium-3, 250 per second, would be injected into a huge combustion chamber and ignited by electron beams. A very powerful magnetic field would confine the nuclear explosions and channel the resulting high-speed plasma out of the rear of the spacecraft. In this way, the pulses of the thermonuclear explosions would provide an enormous thrust and enable Daedalus to reach the desired cruising speed within four years after launch.

The whole vehicle would have a total mass of 54,000 metric tons (119 million pounds), including 50,000 metric tons (110 million pounds) of propellant and 500 metric tons (1,100 thousand pounds) of scientific payload. With a mass almost 20 times that of a fully fueled Saturn V moonrocket, and being propelled by nuclear explosions, Daedalus could not be launched from Earth but would have to be constructed in space. Because the helium-3 required as propellant is very rare on Earth but not on the giant gas planets, Daedalus may even have to be built in orbit around Jupiter.

The first stage of the two-stage ship would be fired for two years to take it to 7.1 percent of the speed of light, and then be jettisoned. Next, the second stage would start and continue to thrust for nearly another two years. Daedalus would then cruise, unpropelled and at a constant velocity of 12 percent the speed of light, for another 46 years before reaching its destination.

At such speed, a collision with even the smallest speck of interstellar dust could be a disaster. For protection, a 7-millimeter (0.3-inch) thick, 50-metric-ton (330,000-pound) disk of beryllium would be placed in front of the spaceship. However, that would not be enough shielding against even the impact of a 1-gram grain of rock, which, at the cruising speed of Daedalus, would hit with the energy of 150 metric tons (330,000 pounds) of TNT explosives.

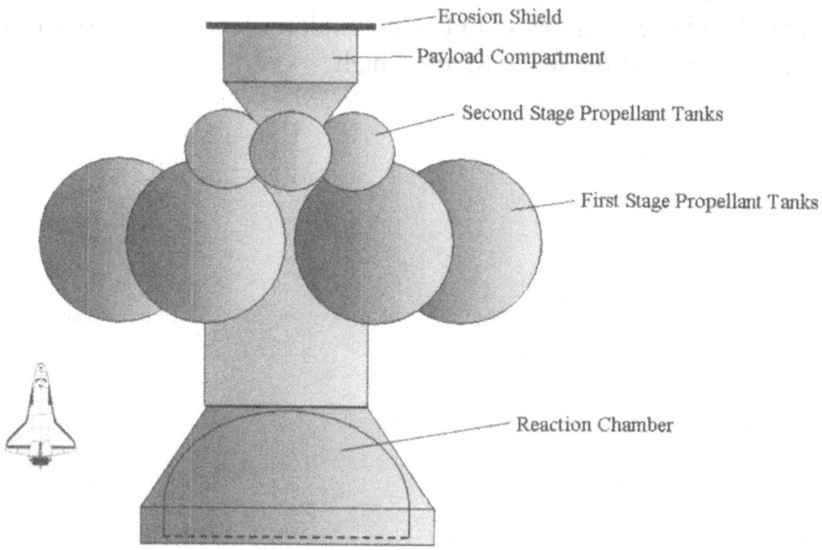

FIGURE 9.6 Layout of the Daedalus concept for an interstellar space probe, with a Space Shuttle Orbiter at the same scale.

Therefore an artificially generated cloud of particles some 200 kilometers (120 miles) ahead of the vehicle would act as an additional shield. Any obstruction would be instantly vaporized by the super-fast impact with these particles. The resulting plasma would be stopped by the beryllium slab (the method is somewhat similar to that of Giotto's Whipple shield described on page 230).

Decelerating the spacecraft would require as much propellant as needed for the departure (and the inclusion of this would mean that even more propellant would be needed to get up to speed in the beginning, due to the higher spacecraft starting mass). However, to limit the size and complexity of Daedalus, its design does not include provisions for deceleration upon arrival.

Daedalus would thus fly by at an incredibly high speed and need to use optical and radio telescopes to investigate Barnard's Star and its hypothetical planets. Some years before arrival it would also launch 18 autonomous probes for more detailed investigations. Powered by nuclear-ion drives, the probes would zoom past any planets and use their suite of instruments especially to look for signs of life and places that look hospitable for biology.

If Daedalus represents one of the more feasible interstellar space probe designs, you can understand why no space agency is planning to build one very soon. The costs for its development and construction would be enormous and far too high for their current budgets. For the foreseeable future, the playground for robotic space explorers ends at the edge of the Solar System. And the fence is rather high.

10

ONLY JUST BEGINNING

OUR exploration and understanding of the Solar System has made giant steps since we first managed to launch space probes into the heavens. Fifty years ago, before the space age, we speculated that we would find soggy swamps on Venus and saw evidence that Mars was covered with vegetation. This invited science fiction writers to populate these planets' surfaces with kings and princesses defending mighty cities against hordes of indigenous monsters, and ancient but desperate civilizations digging vast canals to feed water to their dying planet.

The robotic space explorers we sent to investigate put an end to these romantic visions. Instead of lush jungles with bathing dinosaurs on Venus, they found an overheated and overpressured hell that would be more inspiring to Dante than to famous science fiction writer Edgar Rice Burroughs. Instead of wise aliens living in glass dome cities on Mars, we found a desert planet full of rusty rocks harassed by dry dust storms.

However, we found other and new things to inspire us, worlds we could not have imagined before the space age. We discovered moons with subsurface oceans, methane rain and sulfur volcanoes. We saw brilliant auroras over the poles of giant gas planets and lightning on Jupiter. There are asteroids full of precious metals we may be able to mine some time in the future. We now know that Mars may once have had vast amounts of water running over its surface, and hope that life may still be surviving there somewhere. We developed a new view of Venus that gives us a strong warning about what can be the result of us polluting our own world with greenhouse gasses.

Without actually going out and visiting these alien worlds, we would never have collected more than a tiny fraction of our current knowledge about the planets and their history. When Voyager 1 flew by Saturn in 1980, a NASA JPL scientist commented "We have learned more about the Saturn system in the past week than in the entire span of recorded history."

Even more important, space exploration has given us a new view on ourselves and our place in the Universe. Lunar astronauts showed us the Earth as a shiny blue marble in the infinity of space, a vulnerable small world that has been the set of the amazing evolution of life – life which, after billions of years, has finally developed the ability to reach other planets. Marslanders looked back and saw our planet as just another pale dot in the sky, unremarkable in appearance. If the Earth would suddenly disappear, the rest of the Universe would not notice. Nevertheless, that small world and its intelligent inhabitants have shown the potential to become a Solar System based civilization rather than one stuck on a single planet.

The unique perspective of the living but fragile Earth rising over the dead, gray horizon of the Moon has become a symbol and inspiration for our environmental protection efforts. Space exploration has taught us that our own home planet isn't as infinite and everlasting as humanity subconsciously assumed until the twentieth century.

Earth is just another planet that happened to be in the right orbit around the right star at the right time. It's the only place we know for certain that can support millions of species of life (1.4 million known and between 10 to 100 million unknown species), and it is probably unique. We now know that we have to take care of our own home. Because of all the data gathered by the exploration of the planets and the detailed observations of the Earth, ignorance can no longer be used as an excuse for ruthless exploitation.

Renewed perspectives such as these cannot be expressed in simplistic evaluations of costs and direct financial benefits to society, as some cynical and limited politicians and journalists tend to do. How do you measure the economic benefit of youth being inspired by space exploration to study science and engineering? Or the profit to a society that knows there is more in the Universe than traffic jams, taxes and brainless TV shows? How can we quantify to our descendants the advantages of this generation understanding our planet and its climate a bit better, and therefore leaving them a cleaner, healthier world?

Space exploration has given humanity a context, an improved understanding of what and where we are. Six billion humans on a relatively small planet with a thin and fragile atmosphere, in the middle of an infinite and inhospitable vacuum. You would expect such knowledge to have a uniting and appeasing influence, but with all the wars that occur, the evidence for that is unfortunately rather poor.

Even though the days of newspaper headings shouting "Man walks on Moon" or "Viking 1 lands on Mars!" are over and budgets for space science are more limited than ever, these are exciting times for interplanetary space exploration. During the two years it took to prepare this book, Mars Express and the rovers Spirit and Opportunity arrived at the red planet, and showed us Mars in a way we never saw it before. Then Cassini reached Saturn and deployed the highly successful Huygens probe, which plunged deep into the Titan atmosphere and made it down all the way to the moon's bizarre surface. The ion-propelled SMART 1 went into orbit around the Moon, and is now uncovering new features about our old neighbor in space. Deep Impact was launched and half a year later shot a big hole into an asteroid, and the Japanese Hayabusa probe also reached its target asteroid. The return capsule of the Stardust mission landed back on Earth with a cargo of comet and stellar particles captured in space. We saw the launches of the comet explorer Rosetta, the Mercury orbiter Messenger, the large Mars Reconnaissance Orbiter, Venus Express and the New Horizons mission to Pluto. The Planetary Society's Cosmos 1 experimental solar sail was also launched, but unfortunately crashed due to a malfunctioning rocket engine. Furthermore, NASA announced that it will launch a new probe to Jupiter in 2010, named Juno.

Then there was President Bush's announcement that the future of NASA's crewed spaceflight would be focusing on a return to the Moon and then missions to Mars. The earlier announced Aurora plan of the European Space Agency involves a similar vision, and is also inviting the Russians to be involved.

At the same time, India and China are not only emerging as industrialized countries, but also as eager space-exploring nations. China launched its first astronaut in October 2003 and another two on board a single spacecraft in October 2005. Undoubtedly more will follow, as it has

FIGURE 10.1 ESA and NASA are planning a robotic Mars Sample Return mission to drill up soil from Mars and launch it to Earth in the next decade. Such a mission will also prove technology vital for safely landing and returning future Mars astronauts. [ESA]

plans for space stations and may decide to put its own flag on the Moon in the not too distant future. India now has the capability to launch interplanetary spacecraft with its own rockets, and is planning to send a probe to the Moon.

Curiosity is one of the main defining features of humanity, and exploration is in our blood. We will be sending many more spacecraft to the Moon, Mars and the other planets in the coming decades. Humans may follow but, as always, robotic scouts will lead the way.

No matter how much progress one makes,
There is always the thrill of just beginning.

Robert H. Goddard, rocket pioneer

GLOSSARY

absorption lines: Narrow, dark lines in a spectrum (for instance of the light of the Sun) corresponding to specific frequencies at which radiation is absorbed. They are also called Fraunhofer lines.

aerobraking: Technique consisting of carefully calculated dips into the upper layers of a planet's atmosphere, using the aerodynamic drag to slow down and gradually adjust a spacecraft's orbit.

aerosols: Small droplets floating in an atmosphere.

alpha particle: A subatomic radiation particle consisting of 2 protons and 2 neutrons.

altimeter: An instrument to measure altitude, on spacecraft usually by means of radar or laser, whereby the time it takes a transmitted signal to bounce of a planet's surface and return to the instrument is a measure of the distance to it.

AOCS: The Attitude and Orbit Control System on board a spacecraft.

aphelion: The point in an elliptical, solar orbit where a spacecraft is furthest away from the Sun.

apoapsis: The point in an elliptical orbit where a spacecraft is furthest away from the center of a body being orbited, such as a planet or moon.

ASI: Agenzia Spaziale Italiana – the Italian space agency.

asteroids: Relatively small, rocky bodies that orbit the Sun and which are mainly found between Mars and Jupiter. Like the more icy comets, they are considered left-overs from the formation of the Solar System.

CCD: Charged Coupled Device, consisting of a matrix of detector elements called pixels. Each pixel converts the energy of light particles that hit it into an electrical charge. This electrical charge is subsequently converted into a code that can be stored in a computer memory.

CFRP: Carbon Fiber Reinforced Plastic – a composite plastic material that is reinforced with carbon fibers.

chromatograph: A device able to separate the various components of a mixed gas such as the atmosphere of a planet. Normally used in combination with a mass spectrometer.

CIA: Central Intelligence Agency – the intelligence agency of the United States.

cleanroom: A very clean area in which spacecraft are build.

CNES: Centre National d'Etudes Spatiales – the French space agency.

coarse sun sensor: A sensor that indicates the general direction of the Sun.

comets: Relatively small, icy bodies that orbit the Sun. Like the more rocky asteroids, they are considered left-overs from the formation of the Solar System. The majority exists in vast numbers in the Kuiper Belt and Oort Cloud.

components: The smallest parts from which spacecraft are constructed.

composite: A material that consists of more than one material, combining the favorable characteristics of each.

contractor: A company working on a project under a contract (for instance, from a space agency or a higher level contractor).

control momentum wheel: See "reaction wheel."

corona: The tenuous, outermost layer of the atmosphere of the Sun (or any other star).

cryogenic propellant: A form of rocket propellant that must be kept at extremely low temperatures to remain liquid, such as liquid oxygen (–184 degrees Celsius) and liquid hydrogen (–251 degrees Celsius).

CSA: The Canadian Space Agency

Data Handling: The computer subsystem on board a spacecraft.

DLR: Deutsches Zentrum für Luft- und Raumfahrt – the German space agency.

Doppler effect: The shift in frequency and wavelength of a signal when you move toward or away from its source (or when the source of the signal moves with respect to you). An example is the change in pitch of a train whistle or ambulance siren as it passes by the listener.

Eigenfrequency: A frequency at which an object "likes" to oscillate, also called "natural frequency." When an object is subjected to a vibration with a frequency corresponding to one if its own Eigenfrequencies, it will start to vibrate as well.

electric propulsion: Rocket propulsion based on the expulsion of electrically charged particles.

electromagnetic radiation: Radiation that can be described as consisting of waves of electric and magnetic energy that travel at the speed of light. The entire range of electromagnetic radiation includes radio waves, microwave radiation, infrared, visible light, ultraviolet, X-rays, and gamma-rays.

emission lines: Narrow, bright lines in a spectrum (for instance, of the light of the Sun) corresponding to specific frequencies at which radiation is emitted.

equipment: The individual parts such as sensors, antennas and thrusters out of which a spacecraft is build. The equipment itself consists of smaller parts called components.

ESA: The European Space Agency.

exoplanet: A planet orbiting a star other than the Sun.

Faraday cage: A metal cage or enclosure that prevents electromagnetic radiation such as radio or television signals from entering.

fine sun sensor: A sensors that indicates the precise direction of the Sun.

fluorescence: A molecule absorbing radiation, and then re-emitting it at a lower energy and thus longer wavelength.

Fraunhofer lines: See "absorption lines."

frequency: The number of waves passing a given point in one second, expressed in Hertz (Hz).

gain: The "gain" achieved by an antenna is a measure of the relative amount of incoming radio power it can collect and the strength of the signals it can transmit to Earth. The higher the gain, the stronger the signals it can put out and the weaker the signals it is able to detect.

gamma-ray: The most energetic and shortest-wavelength type of electromagnetic radiation. Gamma-rays have wavelengths less than about 0.1 nanometer.

GEO: Geostationary Earth Orbit – the orbit around the Earth where a spacecraft's orbital rotation is 24 hours and it thus stays over the same point above the equator as the Earth rotates around its axis.

***g*-level:** A measure of the force acting on a body. 1*g* is the force exerted by gravity on the Earth's surface, 2*g* is a force equivalent to twice the weight of an object, etc.

GPS: Global Positioning System – the constellation of satellites that emit signals that make it possible to pinpoint your location on the surface of the Earth.

Gravity Assist Maneuver: A gravity assist maneuver means that a spacecraft swings by a planet and uses that planet's gravitational field and orbital velocity to pick up or lose speed.

gyroscope: Equipment that tells a spacecraft the directions and velocities of its own rotations.

gyroscopic effect: The effect that a spinning object will resist any tilting (due to the so-called momentum inertia of the spinning wheel).

harness: The wires and cables that connect the various equipment on board a spacecraft.

heatpipe: A device that transfers heat by the continuous evaporation and condensation of an internal fluid.

heatshield: A shield that protects a capsule entering an atmosphere at high velocity against the resulting heat.

heliopause: The boundary between the Sun's immense magnetic bubble (called the heliosphere) and the interstellar gas.

heliosphere: The immense bubble surrounding the Sun that is inflated by the solar wind and within which the Sun exerts a magnetic influence.

HGA: High Gain Antenna – an antenna with which relatively strong, narrow beam radio signals can be transmitted and relatively weak radio signals can be received. HGAs enable the use of high data rates.

Hohmann orbit: The most energy efficient orbit for traveling from one planet to another, requiring the lowest speed and therefore the least amount or propellant. In a Hohmann orbit, a spacecraft follows an elliptical orbit that just touches both the orbit of the Earth and that of the target planet.

horizon sensor: A sensor that is able to detect where the horizon is.

ICBM: Inter-Continental Ballistic Missile

IMU: Inertial Measurement Unit, consisting of gyroscopes and accelerometers. The accelerometers measure the rate of change in velocity over a specific period of time, while the gyroscopes measure how fast the spacecraft is turning.

incident light level: The amount of light falling onto an object, which is not only a function of the absolute light level, but also of the angle at which the light hits a surface.

infrared: Electromagnetic radiation of wavelengths longer than the red end of visible light and shorter than microwaves, ranging from about 700 nanometer to 1 millimeter.

ion engine: A form of electric propulsion by which an electric current flows across a magnetic field, creating an electric field directed sideways to the current. When electrically charged atoms, called ions, are put in this electric field, they are accelerated to great speeds, resulting in a form of rocket propulsion.

JAXA: The Japan Aerospace eXploration Agency – the Japanese space agency.

Kuiper Belt: A disk-shaped region past the orbit of Neptune, containing many small icy bodies. It is now considered to be the source of the short-period comets. The planet Pluto is considered to be the largest member of this region. Earth-based telescopes have also found other relatively large planet-like objects in the Kuiper Belt: Quaoar, a planetoid one-third the diameter of the Moon, and Sedna, which has an estimated diameter of about 75% of that of Pluto. Another object, named 2003 UB_{313}, may even be larger than Pluto. Due to their unusual orbits, Neptune's moons Triton and Nereid and Saturn's moon Phoebe are thought to be captured Kuiper Belt objects.

Lagrangian point: A location in space in a rotating two-body system (such as the Earth–Moon system) where the gravitational pulls of these bodies combine to form a point at which a third body of a relatively low mass is stationary relative to the other two bodies. Lagrangian points are named after the Italian-born French mathematician and astronomer Joseph Louis de Lagrange, who first derived their existence.

laser altimeter: An instrument to measure altitude using a laser beam, whereby the time it takes the transmitted signal to bounce off a planet's surface and return to the instrument is a measure of the distance to it.

launch window: The period of time during which a launch can take place, depending on the orbit and position of the target planet, the orbit of the Earth, the Earth's rotation and the trajectory needed for the probe to reach its destination.

LEO: Low Earth Orbit – an orbit between about 100 and 1,500 kilometers altitude.

LGA: Low Gain Antenna – an antenna with which only relatively weak, but very wide beam radio signals can be transmitted and only relatively strong radio signals can be received. LGAs can only be used for low data rates.

light-year: The distance traveled by light in one year. As light moves at 300,000 kilometers (186,000 miles) per second, a light-year is about 9.46 trillion kilometers (5.88 trillion miles).

magnetic field: A field in which magnetic forces are generated by electric currents.

magnetic torquer: Long electromagnets that align themselves with respect to an external magnetic field when electricity is flowing through them. They are used in the attitude control of Earth-orbiting satellites.

magnetometer: A sensor that measures the direction and sometimes the strength of a local magnetic field.

magnetosphere: The region in which a planet's magnetic field dominates that of the solar wind. It has a teardrop shape because on the day side the solar wind pushes on it, drawing out a long magnetotail on the night side.

mass spectrometer: An instrument that can determine the type of gas by measuring the mass of its molecules. Normally used in combination with a chromatograph.

MGA: Medium Gain Antenna – an antenna with which medium strong, medium narrow beam radio signals can be transmitted and medium weak radio signals can be received. MGAs enable the use of medium high data rates.

momentum desaturation/unloading: The periodic slowing down of a reaction (control momentum) wheel, to avoid it becoming useless because it has reached its maximum velocity. Without any compensation, decelerating a wheel would make the spacecraft rotate, but by using opposite thrust from the reaction control system this can be avoided.

Mössbauer effect: See "Mössbauer spectrometer."

Mössbauer spectrometer: A spectrometer using a source of radioactive material to radiate samples with gamma-rays. The spectrometer then analyzes how the radiation gets absorbed by the sample to gain information about its composition (the absorption is called the "Mössbauer effect").

narrow-angle camera: A camera with a relatively narrow field of view, covering less surface area than a wide-angle camera but able to show more detail.

NASA: National Aeronautics and Space Administration – the space agency of the United States.

natural frequency: See "Eigenfrequency."

Oort cloud: An immense spherical cloud of comets that surrounds the Solar System, enveloping the Solar System far beyond the orbit of Neptune. The Oort cloud extends to perhaps around 2 light-years, or 20 million times a million kilometers – half the distance to the next nearest star. The existence of the cloud was first hypothesized by Dutch astronomer Jan Hendrick Oort.

organic molecules: Molecules containing carbon atoms; all known forms of life are based on molecules with chains of carbon atoms, but organic molecules are not necessarily created by life.

oxidizer: The component of rocket propellant that releases oxygen for the combustion of the fuel.

payload: The load carried by a launcher or spacecraft, i.e. the part that is the heart of the mission but not necessary for the basic operation of the vehicle or probe. For a launcher the payload is the spacecraft it carries, for a space probe it is the suite of scientific instruments.

perihelion: The point in an elliptical, solar orbit where a spacecraft is closest to the Sun.

periapsis: The point in an elliptical orbit where a spacecraft is the closest to the center of a body being orbited, such as a planet or moon.

photon: A particle of light that acts as an individual package of energy.

planetary protection: Measures to protect the Earth from contamination by possible microbes from space ("back contamination"), and to protect other bodies in the Solar System from contamination by microbes from Earth carried by visiting space probes ("forward contamination").

planetoids or **planetesimals:** Small, roughly spherical planets that are larger than the often irregularly-formed asteroids, but smaller than the main planets.

plasma: A low-density gas in which some or all of the atoms or molecules are ionized (i.e. electrically charged). The total number of positive and negative charges is equal, therefore the plasma is overall electrically neutral.

Prime Contractor: The main contractor responsible for a space project, reporting directly to the space agency that is financing and organizing the mission.

radiator: Panels that absorb excess heat from inside the spacecraft and radiate it into space.

radio altimeter: An instrument to measure altitude using a radio signal, whereby the time it takes the transmitted signal to bounce off a planet's surface and return to the instrument is a measure of the distance to it.

radiometer: An instrument to detect and measure the intensity of thermal radiation, especially infrared radiation.

reaction wheel: A fairly heavy, solid metal wheel with fixed axes that is used to rotate the spacecraft in different directions using the "action = reaction" principle.

redundancy: Having spare equipment built in that can take over from failing equipment units.

resolution: The size of the smallest objects that can be still be individually distinguished by an instrument (also called "resolving power").

RHU: Radio-isotope Heater Unit, consisting of a few grams of radioactive material contained in protective cladding. A typical RHU emits about 1 watt of heat by the radioactive decay of the material inside.

ring laser gyro: A type of gyroscope that uses the interference pattern between two parts of a split laser beam, caused by the Doppler effect, to measure how fast a spacecraft is rotating.

RTG: Radio-isotope Thermo-electric Generator. RTGs work on the thermo-electric principle that a voltage can be generated between two different conducting materials if they are each kept at different temperatures. In an RTG the temperature difference is created by the heat emitted from the natural radioactive decay of plutonium.

SADM: See "Solar Array Drive Mechanism."

safe mode: The mode of operations to which a spacecraft automatically turns when something goes wrong. In safe mode, the robot puts itself into a stable situation to await orders from Earth.

Sol: A Martian day, which lasts about 24 Earth hours and 37 minutes.

solar array: A panel or set of panels covered with solar cells that generate electrical power from sunlight.

Solar Array Drive Mechanism: A mechanism that keeps a spacecraft's solar array aimed at the Sun by rotating it around its axes.

solar cells: Small, flat cells that convert sunlight into electrical power. Solar arrays on spacecraft typically consist of thousands of interconnected solar cells.

solar wind: A tenuous, radial flow of gas and electrically-charged particles, mostly protons and electrons, streaming from the Sun's outer layer (called the corona). The expanding solar wind drags the solar magnetic field outward. The region of space in which this solar magnetic field is dominant is called the heliosphere.

Solid State Recorder: A type of recorder that, unlike tape recorders and hard disk drives, has no moving parts. Instead it is based on several hundreds of memory chips for storing digital data.

sound barrier: The barrier that was long thought to limit aircraft to subsonic speeds. When an aircraft flies faster than sound, it has broken the sound barrier.

spectrometer/spectroscope: An instrument used for splitting light or other forms of electromagnetic radiation into the component wavelengths, called a spectrum. If a camera or detector is used to record the spectrum, the device is known as a spectrograph.

spectrograph: See "spectrometer/spectroscope."

spectrum: A plot of the intensity of electromagnetic radiation at different wavelengths. In the case of visible light, the spectrum appears as a rainbow of colors (in fact, a rainbow is itself a spectrum of the Sun caused when its light is refracted by raindrops).

spin stabilized: Spacecraft stabilization by rotation; as long as the probe is in balance, the gyroscopic effect keeps the rotational axis in the same direction.

star mapper: An optical sensor that follows several stars in its field of view (as opposed to star trackers, which follow only one star). By remembering the positions of the stars with respect to each other, its electronics can determine the orientation of a spacecraft in all three dimensions.

star tracker: An optical sensor that is pointed at a single star. By tracking how this star is moving in the sensor's field of view, a star tracker can tell a spacecraft the direction in which it is moving or rotating. By keeping a star tracker trained at a certain star, the side of the space probe on which it is located is always pointed in the same direction.

stereoscopic: Combining two pictures made from slightly different angles, a three-dimensional view can be created. Because people have two eyes observing objects at different angles, we have stereoscopic vision.

subcontractor: A company working on a project under a contract with another company (a higher level contractor).

subsystem: An assembly of equipment that work together to perform a particular task on board a spacecraft. Examples are the Thermal Control subsystem and the Power subsystem.

sun sensor: A sensor that indicates the direction of the Sun (see also "coarse sun sensor" and "fine sun sensor").

three-axis stabilized: The spacecraft's attitude is actively controlled around all three possible rotational axes.

trans-Neptunian objects: Objects orbiting beyond Neptune, including Pluto.

ultraviolet: Ultraviolet (UV) light consists of electromagnetic radiation with wavelengths shorter than those of visible light, but longer than those of X-rays. UV wavelengths range from 100 to 350 nanometers, while Extreme Ultraviolet (EUV) ranges from 10 to 100 nanometers.

Van Allen Belts: Two doughnut-shaped belts of high-energy charged particles that are trapped in the Earth's magnetic field. They were discovered in 1958 by James Van Allen using measurements made by the Explorer 1 satellite. The inner Van Allen Belt lies about 9,400 kilometers above the equator, and contains protons and electrons from both the solar wind and the Earth's ionosphere. The outer belt is about three times further away and contains mainly electrons from the solar wind.

wavelength: The wavelength of electromagnetic radiation is the distance between any two corresponding points on successive waves (for instance, from one peak to the next).

Whipple shield: An impact shield consisting of two protective layers. The first vaporizes all but the largest of incoming particles, while the second absorbs any debris that pierced the front barrier. With two relatively thin layers, a Whipple shield can protect against particles that otherwise could only be stopped with a much thicker and heavier single layer shield.

wide-angle camera: A camera with a relatively wide field of view, covering more surface area than narrow-angle cameras but not able to show as much detail.

X-ray: Electromagnetic radiation with wavelengths shorter than those of ultraviolet rays and longer than those of gamma-rays. X-ray wavelengths range from one hundredth of a nanometer to 10 nanometers.

BIBLIOGRAPHY

Books and Reports

Battrick, Bruce. *BepiColombo, Interdisciplinary Mission to Planet Mercury*. ESA brochure BR-165, ESA, The Netherlands, 2000.

Calder, Nigel. *Beyond This World*. ESA Publications Division, Noordwijk, The Netherlands, 1995.

Calder, Nigel. *Giotto to the Comets*. Presswork, London, UK, 1992.

Cooper, Henry S.F. Jr. *The Evening Star; Venus Observed*. Farrar Straus Giroux, New York, USA, 1993.

Couper, Heather and Nigel Henbest. *The Planets*. Pan Books Ltd, London, UK, 1985.

Dyson, George. *Project Orion: The True Story of the Atomic Spaceship*. Henry Holt & Company, Inc., USA, 2002.

Evans, Ben with David M. Harland. *NASA's Voyager Missions: Exploring the Outer Solar System and Beyond*. Springer–Praxis Books, London, UK, 2004.

Fortescue, Peter and John Stark. *Spacecraft Systems Engineering* (Second Edition). John Wiley & Sons Ltd, UK, 1995.

Fortey, Richard. *Trilobite! Eyewitness to Evolution.* Flamingo, London, UK, 2000.

Gardiner, Brian, Barry Cox, R.J.G. Savage and Dougal Nixon. *The Macmillan Illustrated Encyclopedia of Dinosaurs and Prehistoric Animals.* Marshall Editions Limited, UK, 1988.

Gatland, Kenneth. *The Illustrated Encyclopedia of Space Technology.* Salamander Books Ltd, London, UK, 1981.

Hogg, Ian V. *German Secret Weapons of the Second World War; The Missiles, Rockets, Weapons and New Technology of the Third Reich.* Greenhill Books, London, UK, 2002.

Houston, A. and M. Rycroft. *Keys to Space: An Interdisciplinary Approach to Space Studies.* International Space University, McGraw-Hill, USA, 1999.

Leyder, Jean-Christophe. *Analysis of Messenger, the NASA Mission to Mercury.* ESA and Université de Liège, The Netherlands, 2004.

Matson, Wayne R. *Cosmonautics, a Colorful History.* Cosmos Books, Washington, USA, 1994.

McElyea, T. *A Vision of Future Space Transportation; A Visual Guide to Future Spacecraft Concepts.* Apogee Books, Burlington, Canada, 2003.

McNab, David and James Younger. *The Planets.* BBC Worldwide Ltd, London, UK, 1999.

Moore, Patrick. *The Guinness Book of Astronomy Facts and Feats.* Guinness Superlatives Ltd, UK, 1984.

NASA Institute for Advanced Concepts. *NIAC Annual Report July 2004.* NIAC, Atlanta, USA, 2004.

NASA JPL. *NASA Facts: Viking Mission to Mars.* NASA JPL, Pasadena, USA, 1988.

NASA JPL. *NASA Facts: Cassini Mission to Saturn.* NASA JPL, Pasadena, USA, 2003.

NASA JPL. *Mars Exploration Rover Landings Press Kit.* NASA, Pasadena, USA, 2004.

Rich, Ben R. and Leo Janos. *Skunk Works.* Warner Books, London, UK, 1995.

Smolders, Piet. *De zwaartekracht voorbij, veertig jaar ruimtevaart.* Schuyt & Co., Haarlem, The Netherlands, 1997.

Sykes, J.B. *The Other Side of the Moon.* Pergamon Press, Oxford, UK, 1960.

Turnill, Reginald. *The Observer's Book of Unmanned Spaceflight.* Frederick Warne & Co. Ltd, London, UK, 1974.

Ulivi, Paolo with David M. Harland. *Lunar Exploration, Human Pioneers and Robotic Surveyors.* Springer–Praxis Books, London, UK, 2004.

Zimmerman, Robert. *Genesis, the Story of Apollo 8*. Dell Publishing, New York, USA, 1999.

PAPERS AND ARTICLES

Atzei, Alessandro. "De zoektocht naar leven op Europa," *Ruimtevaart* (February), The Netherlands, 2005.

Bearden, David A. "Small-satellite costs," *Crosslink*, Winter 2000/2001, USA, 2000.

Chang, Kenneth. "Not science fiction: An elevator to space," *New York Times* (September 23), USA, 2003.

Christensen, Philip R. "The many faces of Mars," *Scientific American* (July), USA, 2005.

Colozza, Anthony. "Feasibility of a long duration solar powered aircraft on Venus," *AIAA* 2004-5528, 2004.

Day, Dwayne A. "Those magnificent spooks and their spying machines," *Spaceflight*, vol. 39, no. 3, UK, 1997.

Evans, Nigel. "Cassini launch," *Spaceflight*, vol. 40, no. 2, UK, 1998.

ESA Press Release. "Mission to Mars set to revolutionise ESA's working methods," *ESA Press Release,* number 12-1999, 1999.

Falkner, P. "Update on ESA's Technology Reference Studies," *56th International Astronautical Congress*, Fukuoka, Japan, 2005.

Furniss, Tim. "Drawing blunder may be to blame for Genesis crash," *Spaceflight*, vol. 46, no. 12, UK, 2004.

Haar, Gerard van de. "Messenger on its way to Mercury," *Spaceflight*, vol. 46, no. 10, UK, 2004.

Holt, Roger. "A brief history of cometary exploration," *Spaceflight*, vol. 47, no. 9, UK, 2005.

Hillebrand, J. and F.J.P. Wokke. "Robots op ontdekkingsreis," *Ruimtevaart* (June), The Netherlands, 1998.

Landis, Geoffrey. "Pathfinder – a personal retrospective," *Spaceflight*, vol. 44, no. 8, UK, 2002.

Merkle, R.C. "Self replicating systems and molecular manufacturing," *JBIS*, vol. 45, no. 10, UK, 1992.

van Pelt, Michel. "Big future for small spacecraft," *Spaceflight*, vol. 40, no. 3, UK, 1998.

Powell, Joel W. "The forgotten mission of Pioneer 5," *Spaceflight*, vol. 47, no. 5, UK, 2005.

Sanders, H.M. and C.M. Wentzel. "Onconventionele lanceersystemen:

schieten met satellieten," *Ruimtevaart* (February), The Netherlands, 2000.
Schilling, Govert. "De killer moet een duwtje krijgen," *De Volkskrant* (June 18), The Netherlands, 2005.
Schilling, Govert. "We gaan weer eens naar de hel," *De Volkskrant* (October 22), The Netherlands, 2005.
Scala, Keith J. "A history of nuclear space accidents," *Spaceflight*, vol. 40, no. 4, UK, 1998.
Silvester, John. "Manned space flight in the sidings," *Spaceflight*, vol. 45, no. 10, UK, 2003.
Simpson, Clive. "Smash and grab, comet mission opens up Solar System secrets," *Spaceflight*, vol. 47, no. 9, UK, 2005.
Suriano, Robyn and Tony Boylan. "Demonstration ends with 27 arrested (on anti-Cassini demonstration)," *Florida Today* (October 5), USA, 1997.
US Department of Energy. *Department of Energy Facts: Radioisotope Heater Units.* US Department of Energy, USA, 1998.

WEBPAGES

Asaravala, Amit. "Congressman backs Asteroid Agency": www.wired.com/news/space/0,2697,67697,00.html, 2005.
Bedway, Barbara. "Rocks roll in from outer space!": teacher.scholastic.com/researchtools/articlearchives/space/rockroll.htm, 2005.
Berger, Brian. "New Horizons set to launch with minimum amount of plutonium": www.space.com/spacenews/businessmonday_041004.html, 2005.
Cassini-Huygens Tracking Data: saturn.jpl.nasa.gov/mission/nav.cfm, 2005.
China in Space website: www.spacetoday.org/China/China.html, 2004.
Computer Hope.com computer history: www.computerhope.com/history, 2005.
Daedalus project summary: www.daviddarling.info/encyclopedia/D/Daedalus.html, 2005.
David, Leonard. "Mars rover: digging out of tough terrain": www.space.com/missionlaunches/050509_opportunity_sand.html, 2005.

Digital Learning Center for Microbial Ecology: Microbe Zoo; Space Adventure: Martian Bacillus?: commtechlab.msu.edu/sites/dlc-me/zoo/zsa0900.html, 2004.
Dunkle, Terry. "The big glass": www.terrydunkle.com/glass.htm, 2005.
"Eight-month mission lasts fourteen years": www.boeing.com/defense-space/space/bss/factsheets/scientific/pioneer/pioneer.html, 2005.
The Electromagnetic Spectrum (NASA student features): www.nasa.gov/audience/forstudents/5-8/features/F_The_Electromagnetic_Spectrum.html, 2005.
ESA Venus Express website: www.esa.int/SPECIALS/Venus_Express/index.html, 2005.
European Space Agency main website: www.ESA.int, 2005.
Evans, Ben. "The Galileo trials": www.spaceflightnow.com/galileo/030921galileohistory.html, 2005.
Hamilton, Rosanna L. "The Oort Cloud": www.solarviews.com/eng/oort.htm, 2005.
Hayabusa/Muses-C: www.isas.ac.jp/e/enterp/missions/hayabusa, 2005.
History of Space Exploration: www.solarviews.com/eng/history.htm, 2005.
How Stuff Works website: www.howstuffworks.com, 2005.
India in Space website: www.spacetoday.org/India/India.html, 2004.
Interplanetary Society Cosmos 1 solar sail website: www.planetary.org/solarsail, 2005.
IRDT (Inflatable Re-entry and Descent Technology): www.esa.int/SPECIALS/ESA_Permanent_Mission_in_Russia/SEMOR61XDYD_0.html, 2005.
"Japan's asteroid probe to head home despite glitch": www.space.com/missionlaunches/ap_051127_hayabusa_update.html, 2005.
Jet Propulsion Laboratory missions website: www.jpl.nasa.gov/missions, 2005.
"Mars Express contract signed": www.marssociety.org/bulletins/bulletin_200499_01d.asp, 2005.

Martinez, Carolina. "Unusual geology seen during Enceladus fly-by": www.esa.int/SPECIALS/Cassini-Huygens/SEMVTX808BE_0.html, 2005.
Mars Global Surveyor Mission Overview: www.marsnews.com/missions/mgs, 2005.
Mars Reconnaissance Orbiter website: marsprogram.jpl.nasa.gov/mro, 2005.
National Science Foundation Press Release. "Astronomers announce the most Earth-like planet yet found outside the Solar System": www.nsf.gov/news/news_summ.jsp?cntn_id=104243&org=NSF&from=news, 2005.
Nakhla Dog Meteorites: Mbale L5-6 Chondrite: www.nakhladogmeteorites.com/catalog/mbale.htm
NASA National Space Science Data Center (catalog of spacecraft): nssdc.gsfc.nasa.gov/planetary/projects.html, 2005.
NASA main website: www.nasa.gov, 2005.
NASA's Basics of Spaceflight website: www2.jpl.nasa.gov/basics, 2005.
National Air and Space Museum: Exploring the Planets: www.nasm.si.edu/research/ceps/etp/etp.htm, 2005.
Nuclear Power in Space: www.nuc.umr.edu/nuclear_facts/spacepower/spacepower.html, 2005.
Nuclear Space, the Pro-Nuclear Space Movement: www.nuclearspace.com, 2005.
RussianSpaceWeb.com website about Baikonur cosmodrome: www.russianspaceweb.com/baikonur.html, 2004.
Silber, Kenneth. "Cassini's flyby: getting a gravity boost": www.space.com/scienceastronomy/solarsystem/gravity_assist.html, 1999.
Spacefaring Japan website: www.spacetoday.org/Japan/Japan/History.html, 2004.
Space Today website on the Solar System: www.spacetoday.org/SolarSystem.html, 2005.
Trailblazer website: www.transorbital.net, 2005.
Universe Today. "SOHO nears 1,000th comet discovery": www.universetoday.com/am/publish/soho_1000_comets.html
University of Bremen Robotics Group homepage: www.informatik.uni-bremen.de/robotik, 2005.

Wade, Mark. *Encyclopedia Astronautica*: www.astronautix.com, 2005.
Wilson, Jim. "Galileo's antenna: The Anomaly at 37 million miles" (JPL Universe 7/3/92): www2.jpl.nasa.gov/galileo/anomaly.html, 2005
Zak, Anatoly. "Venera-7, Survivor of Venusian crash" www.space.com/news/spacehistory/venera7_000817.html, 2005.

OTHER

Sierhuis, Maarten. "Praten met Robots," Presentation for *Mars, van Robots naar Mensen* symposium, Dutch Mars Society, Amsterdam, 2005.

INDEX

Illustrations are indicated by bold italic figures

Printed in the United States of America.